Designing Usable Products for African Users

Designing Usable Products for African Users: The African Product Resilience (APR) Framework™ dismantles the flawed practice of copying Western design models for African markets. The book introduces the African Product Resilience (APR) Framework™, providing actionable steps to engineer technology specifically for adversity. It includes real-world case studies, expert insights, and practical checklists covering data efficiency, offline functionality, and hyper-local customization. The framework highlights four core pillars: digital access, trust, digital literacy, and sociocultural context.

The target readers are founders, product managers, designers, researchers, software engineers, and global strategy teams aiming to properly reach the continent and its people. It also serves investors and VCs evaluating a startup's market fit and long-term sustainability in Africa.

Readers will learn how to create resilient, inclusive products that thrive despite severe infrastructural constraints, such as expensive data, inconsistent power, and low-end shared devices. They will discover how to earn user trust in low-trust environments and avoid costly product failures caused by cultural blind spots. Ultimately, the book empowers tech teams to scale efficiently across diverse African markets without wasting resources, turning systemic constraints into competitive advantages.

"This book delivers the blueprint the African tech ecosystem has been waiting for. It defines the standards for efficiency, reliability and usability required to serve the 'next billion users,' and is essential for founders and investors who want to make impact and build resilient businesses."
Tosin Eniolorunda, *Founder & Group CEO, Moniepoint Inc.*

"Finally, a guide that matches the reality we've seen doing research in Africa over the last decade. *Designing Usable Products for African Users* offers a practical framework to successfully navigate the continent's unique infrastructure and cultural landscape."
Yann Le Beux, *Co-Founder & CEO, YUX Design*

Designing Usable Products for African Users

The African Product Resilience (APR) Framework™

Iwalola Sobowale, Lade Tawak,
and Hillary Omitogun

Illustrated by Akachi Ogbonna

Editor: Oluwadunni Adegbite

CRC Press
Taylor & Francis Group
Boca Raton London New York

CRC Press is an imprint of the
Taylor & Francis Group, an **informa** business

Designed cover image: Akachinonyerem Ogbonna

First edition published 2027
by CRC Press
2385 NW Executive Center Drive, Suite 320, Boca Raton FL 33431

and by CRC Press
4 Park Square, Milton Park, Abingdon, Oxon, OX14 4RN

CRC Press is an imprint of Taylor & Francis Group, LLC

ISBN: 978-1-041-09044-1 (hbk)
ISBN: 978-1-041-08988-9 (pbk)
ISBN: 978-1-003-64819-2 (ebk)

DOI: 10.1201/9781003648192

Typeset in Minion
by KnowledgeWorks Global Ltd.

Contents

PART 3 Looking Forward

Foreword

There is a sentence in this book that stayed with me long after I read it: Africa is the proving ground for truly adaptable and user-centric technology. That line captures what many of us who build here have learned the hard way: if your product survives here, it can survive anywhere.

And if you want it to thrive, you must design with honesty and intentionality that acknowledges the lived realities of African users, not the imagined ones. This book gives us the language, the tools, and the courage to do that.

I have spent the past decade operating in Nigerian fintech, navigating regulatory chaos, designing for multiple versions of low connectivity, and dealing with customers whose trust we must earn every single day. I've also learned that product teams elsewhere often underestimate what it takes to build for this continent. They assume the challenge is scale, or language, or infrastructure. But the real challenge is context. The real challenge is designing for a user navigating systemic uncertainty while trying to get a small digital task done without losing money, time, or dignity.

Designing Usable Products for African Users: The African Product Resilience (APR) Framework™ meets that reality head-on. It is both deeply practical and deeply empathetic, a combination that I believe that we desperately need.

Africa is not an emerging version of somewhere else.

One of the first gifts of this book is its dismantling of the idea that Africa is simply "a developing market" waiting to catch up. It argues instead that the constraints here—power instability, expensive data, low-end devices, phone-sharing, fragmented cultures—force a level of product resilience that the rest of the world is only now beginning to value. And they are right.

Products built for our "edge cases" often end up being more universal, more inclusive, and more durable. Whether it's designing for 2G, supporting USSD, enabling offline mode, or creating flows that work seamlessly through intermittent connectivity, these are not compromises. These are competitive advantages. They are the foundation of what the authors have brilliantly called the African Product Resilience (APR) Framework™, a structured way to build technology that thrives in adversity.

Designing for real life, not ideal conditions.

One of my favorite parts of this body of work is how unapologetically it foregrounds the user's real environment. Not the environment we wish existed, not the one shown in pitch decks, but the one people actually live in.

And that environment is messy.

Here, usability is not a question of whether an interface is pretty. It's whether a user can complete a task before their data finishes, their battery dies, their network drops, or their shared device (yes, we still have those) is reclaimed by the next person. It is whether a product respects the fact that a user may toggle off mobile data every 10 minutes to conserve costs, or that they may be literate in English but not fluent in it, or that they trust their neighbor's recommendation more than your very smart onboarding copy.

Trust is not an add-on. It is the product.

Anyone who has worked in African fintech will tell you: before users ask, "Is this easy?", they ask, "Is this safe?" Our true currency is trust. And the authors treat it with the seriousness it deserves.

This book shows us that trust in African digital products is layered. Built on regulation, cultural norms, visual cues, human validation, transparent microcopy, and a product's ability to never surprise users negatively. Something as simple as clearly indicating data usage, or explaining why an app is asking for permissions, or designing around the actual realities, can be the difference between adoption and abandonment. It was refreshing to see such nuance reflected here.

Culture is not an aesthetic. It is a system.

I appreciated how this book treats culture as a structural force rather than a decorative detail. It explains, with clarity and research, how cultural dimensions like power distance, collectivism, uncertainty avoidance, and communication styles shape user behavior. And it warns against the lazy generalizations that often plague "African markets."

What it offers instead is a more mature lens: *culturalization*, not just localization. That means building features that respect real behaviors like haggling, multiple SIM use, mixed payment preferences, or security concerns tied to gender and multiple users using the same device. It means designing for the social realities in which products live.

Research must be as local as the problems we're solving.

The chapters on user research are among the most important in this book. They confirm a truth many of us have lived: conventional research methods simply do not work when participants have limited literacy, unpredictable network access, or deep social reasons to withhold honesty. The authors outline techniques that actually function here: community-based recruiting, participatory research, indirect elicitation, and enumerator-led surveys, and they do so with clarity and respect for the user's humanity.

Research, they remind us, is not just about discovering insight; it is about earning the right to be told the truth.

A future worth building.

Perhaps the best thing about this book is that it is not pessimistic. It is refreshingly grounded, but also hopeful. It sees Africa not as a problem to be solved, but as a frontier that will shape the future of global design. It highlights a coming world where hybrid digital-physical systems, privacy by design, low-resource optimization, multilingual AI, and culturally authentic products become the global standard.

And honestly, I believe that too. The next generation of global design principles will come from here, from home.

From the teams who have been forced to design for complexity, fragmentation, low literacy, and low trust.

From the people who have built products that run reliably on the worst networks, who have designed interfaces that work with or without text, and who have kept millions of users safe despite infrastructure that fails them daily.

A book for builders, created by builders.

This is not a theoretical work. It is not written from the outside looking in. It is built on field research, lived experience, product failures, regulatory surprises, culture-specific insight, and a deep respect for the user. It is written by practitioners who have done the work, felt the friction, and learned the lessons. I should know, I have seen the authors at work.

But most importantly, it is written with empathy. Not just for the user but also for the builders: the designer, PM, engineer, founder, or researcher trying to create something that lasts in a challenging environment.

As someone who has worked in this ecosystem for a long time, I recognize myself in these pages. I recognize many of the mistakes, many of the constraints, and many of the quiet triumphs.

This book is a guide, but it is also a mirror. And I believe it will become foundational reading for anyone serious about building technology for Africans and, eventually, for everyone else too.

Odunayo Eweniyi
Co-Founder & COO, PiggyVest

Preface

Why we need an "African Usability Bible"

Product design in Africa has too often been treated as a simple adaptation of models developed elsewhere—models that came from completely different socioeconomic, cultural, and infrastructural realities. This approach has led to poor user experiences at best, and major product failures at worst. Despite these shortcomings, there remains a significant literature gap on building digital products in and for Africa.

Western-centric design principles are glaringly inadequate for the African context because they rest on faulty assumptions about users, infrastructure, and access.

Consider the common assumption of widespread, high-speed internet and consistent electricity. In many parts of Africa, this is not the reality. So a product designed without robust offline capabilities or data efficiency will quickly fail its intended users. Similarly, most digital tools assume a high level of digital literacy. For many African users, a mobile phone may be their first and only computer.

One clear example of this failure to adapt design to African realities is in attempts to copy complex e-commerce platforms without considering local payment preferences. Credit card payments are standard in many Western markets, yet cash remains a common payment method across much of Africa. This is due to a lack of trust in online transactions, limited access to formal banking, and a preference for immediate satisfaction. A platform that does not support cash payments or makes them difficult to use will create friction for users.

All these factors highlight the need for democratized insights on the realities of building for the continent. As such, this book will make practical, African-specific usability knowledge accessible.

Investing in and building African tech requires a deep understanding of the human and market context. Truly great products are built on a foundation of deep understanding aimed at serving the user, anticipating their challenges, and integrating smoothly into their daily lives.

While the heavy lifting of closing coverage gaps and solving systemic infrastructure problems lies with government bodies and industry stakeholders, this book serves as a critical evidence base for those decisions. Policy cannot be effective if it is designed for an idealized version of Africa. By documenting lived realities, we hope to offer stakeholders a roadmap for where intervention is most needed.

Beyond guiding builders, this book aims to highlight the friction points that policymakers must address to unlock the digital economy. We hope to provide a clear picture of the constraints users face, helping both builders develop resilient solutions and regulators craft policies that protect users without stifling the innovation that keeps them connected.

Who is this book for?

This book is for those who are tired of copy-pasting global solutions that fail when they meet the reality of the African market. Specifically, this book is for:

1. Founders and Product Managers looking for a framework to scale products across diverse markets without losing user trust or burning resources
2. Designers and Researchers who realize that "standard" design patterns are failing their users and need a new set of heuristics for our environments

3. Software Engineers who want to understand the hardware and infrastructure constraints users face in accessing the solutions they're building
4. Investors and VCs who want to better understand the African market and the factors that enable a startup's long-term sustainability and market fit
5. Global Strategy Teams at international tech companies seeking to reach the "Next Billion Users" by understanding the nuances of the continent

Let's figure out if this book is for you: Do you agree or disagree with the following assumptions?

	Assumptions	Agree or Disagree?
1	If we design for mobile first, we are covering the African market	
2	We should wait for better internet and power grids before launching complex services	
3	Translating an app into Swahili or Yoruba is enough to "localize" the product	
4	Showing a "Verified by Visa" logo or an SSL badge will make users trust your platform	
5	Since data is expensive and income is low, offering a service for free will guarantee adoption	
6	We must build an app that the farmer or the market trader can use directly	
7	Facial recognition and image algorithms developed globally will work fine once deployed in Africa.	
8	We should move users away from "clunky" USSD codes toward modern, sleek app interfaces.	
9	Beautiful, high-resolution imagery and animations make an app feel "premium" and trustworthy.	
10	A fintech app that works in Lagos (Nigeria) will work in Nairobi (Kenya) with minimal changes.	
11	To scale, we must make onboarding 100% digital and automated.	
12	Our app should always provide the most up-to-the-second data and location tracking	

(Continued)

	Assumptions	Agree or Disagree?
13	We should use AI-driven Natural Language Processing to support local dialects	
14	African users care only about functionality—data privacy or regulatory compliance are not as important	
15	Designers and builders educated at top global institutions are best equipped to build African products	
16	Facebook and Google ads are the fastest way to drive user stickiness.	
17	If a user can't navigate our flow, they just need more "education" on how to use tech	
18	To truly scale across the continent, we must eliminate human intermediaries from our processes	
19	Every user has their own private device protected by their own biometric lock	
20	A user is either "online" or "offline"	

If you agree with many of the above statements or sit on the fence with your answer, then this book is for you. We are challenging the failed assumptions that have led to product and design decisions in Africa.

A note from Iwalola

Every day, across Africa, brilliant digital ideas that aim to solve real problems struggle for adoption.

I see it every day, and you probably do too. Apps, websites, and digital solutions that launch with the promise of making life easier, connecting communities, or boosting businesses, yet fall short. Why? They're complex. They drain data. Or they simply don't fit into the rhythm of daily life for the people they're meant to serve. Nobody intends to build a bad product, but it happens time and again—from ambitious startups to established players, across industries and throughout our vibrant tech ecosystem.

It begs the question: Why are we still missing the mark so often?

For too long, the conversation around product design and user experience has been dominated by perspectives and principles born in different worlds, for different realities. We've learned the frameworks, mastered the tools, and absorbed the jargon. But what happens when those well-intentioned lessons collide with Africa's unique and incredibly diverse sociocultural landscape? What happens when limited bandwidth is a fundamental functional reality, or when cultural cues, subtle to an outsider, are critical for building trust?

The challenge is moving beyond functional products to creating digital experiences that genuinely understand, empower, and resonate with African users. Technology has brought millions online and thrown open the doors to unprecedented opportunities, transforming lives across our continent. But access alone isn't enough. We're witnessing an explosion of innovation—from groundbreaking fintech solutions like M-Pesa in Kenya, which revolutionized mobile money, to e-commerce giants adapting logistics for complex urban landscapes, and edtech platforms bridging educational gaps. The potential is immense, but realizing it demands a deep, localized understanding of who we are designing for.

This book is your guide.

We won't rehash generalized principles. Instead, we'll show you how to tailor them, twist them, and sometimes toss them aside to build truly usable and resilient products for African users. We'll explore how to move beyond merely building "functional" solutions to designing for the incredible, diverse, and rapidly evolving market on our continent. From understanding users juggling multiple SIM cards and inconsistent power to navigating the subtle yet profound influences of community and language, you'll gain the insights, frameworks, and practical strategies needed to create products for a thriving demographic. Consider this your essential daily resource for keying into the full potential of African digital innovation.

A note from Lade

I'm a firm believer that ignoring the needs of so-called "edge cases" leads to frustrating, poorly designed experiences—not just for those "edge cases", but for everyone eventually. Designing for those considered "out of scope" often ultimately leads to a more inclusive web, a better experience for all users, and innovative business solutions.

Though the stories and lessons here focus primarily on building with and for Africans, the principles and benefits extend beyond the continent. Limitations in technological access and digital literacy are global issues, and designing for them improves experiences for all users. We saw this during the global lockdowns brought on by COVID-19, when people all over the world struggled to continue participating in the systems they had previously been part of because of poor or limited internet connectivity at home, even in "first-world" countries.

Having contributed to building many products that serve users across the globe, I have seen firsthand how the things many builders take for granted can make or break the user experience. I once witnessed first-time smartphone users in a research study struggle with seemingly "simple" tasks like adding a contact or downloading an app. The more I did research, the more I noticed these constraints and experiences, which I have referred to elsewhere as "technological accessibility."

In conversations with Iwalola and Hillary, we realized we had similar ideas about what it takes to build truly usable and successful products with and for Africans, and that led to coauthoring this book. Ultimately, I hope this book serves as a starting point for more work on understanding how to build solutions that actually work *with and for* a growing continent.

A note from Hillary

I have always believed that the best way to move beyond surface-level knowledge into real expertise is to write extensively about what I'm still working to understand. Writing forces me to sit with the unknowns and turn messy, lived experiences into ideas that others can actually use.

So when Lade, Iwalola, and I realized we had all been observing similar gaps between how products are designed and how they are actually used—and the potential that emerges when we design from these realities—writing a book felt like a natural next step.

Throughout my work as a researcher, I've been most drawn to the moments when things don't go as planned. When data from Addis Ababa forced me to rethink everything I thought I understood about how people navigate interfaces under constraint. When a user in Nairobi explained a workaround so elegant that it revealed how fundamentally my questions had missed the point. These moments, uncomfortable as they are, are where real learning happens.

They've also taught me something important: good research begins with accepting uncertainty and learning to design around it. Working across African markets has shown me how powerful it can be when teams listen deeply, adapt their methods, and shape products around real life rather than expecting real life to adapt to the products.

This book is the result of years spent learning to listen, adapt, and rethink how research and product development happen in complex, fast-changing environments. I hope it helps teams embrace uncertainty with curiosity and humility, and gives you the confidence to design for real life as it is.

Acknowledgements

Names in alphabetical order

We owe a great deal of gratitude to several people whose help and support made this book possible.

Thanks first to our project manager, *Joy Aregbesola,* and to the volunteers who helped recruit and interview experts and gather case studies:

Abdou Berekou

Emmanuella Aiyeola

Hairat Alatishe

Odunayo Akinlade

Sinon Aifuwa

To all the experts who contributed through interviews or written submissions:

'Tife Ekundayo, Managing Partner, Nexa Advisory

Adeola Adeyemi, Product Manager, M-Kopa

Aderinsola Adejuwon, Product Designer

Adewale Adeyemo, Founder

Adewunmi Aladenusi, Senior Product Designer, Flutterwave

Akintomiwa Olaoluwa, Product Designer

Akinyi Odera, Project Director, Data Use Partnership for Ethiopia—Think Place

Anukriti Kedia, Head of UX, Moniepoint
Ayomide Ekerin, Head Growth Marketing, 9PSB
Beryl Njoki, Founder, New Beginnings Kwa Ground (NBKG)
Boluwatife Sonde, Product Designer
Bridget Iyere, Lead Product and Operations, eLimu world
Busolami Kehinde, Product Manager
Christopher Ottobong, Product Designer
Ebuka Arinze, Product Manager
Femi Hanson, Head of Marketing, Palmpay
Jerahmeel Madumere, Information Security Researcher
Joseph Brendan, Developer
Joshua Adenuga, Head of Product, FPG Technologies
Kome Sideso, Deputy Head, CcHUB Design Lab
Lilian Ekpere, Product Manager
Mutugi Kirema, Senior UX Researcher, Moniepoint
Ndubuisi Anyaoku, Head of Product Design (Loyalty), Revolut
Olabinjo Adeniran, Product and Growth Marketing Consultant
Oluchi Audu, Senior Design Researcher, YUX Design
Oluwapelumi Oyetimein, Head of Growth
Ope Adeyemi, SVP, Onboarding and Retention, Moniepoint
Paul Otu, Head of Design Strategy and Practice, Interswitch
Princess Akari, Product Manager, Moniepoint
Réné Ahonbo, Product Manager
Rukayat Yaro, Product Designer
Seyi Oniyitan, Product Design Lead, Rise
Susan Onigbinde, Founder, DODO Design, Africa
Theophillus James, Product Manager
Tochy Emereole, API Marketing Lead, Quidax
Toyib Mustapha, Software Developer
Tunde Mason, Head of Product Design, Kuda
Yann Le Beux, Co-Founder & CEO, YUX Design

And to those who contributed their time, skills, and support by providing information, giving us feedback, and making introductions to help us bring this book to life:

Akachi Ogbonna, Designer & Illustrator
Nnanna Ijezie, Product Manager
Odun Eweniyi, Co-Founder & COO, PiggyVest
Oluwadunni Adegbite, Editor
Tosin Eniolorunda, Founder & Group CEO, Moniepoint Inc.

About the Authors

Iwalola Sobowale is an executive leader in Research and Strategy, and founder of Usability for Africa. With 13+ years bridging customer insights and business value, she builds high-growth products through embedded research. A speaker at *World Usability Day Africa 2024* and *ESOMAR Africa in Focus 2026*, her publications include "From NPS to Action" and *Designing Usable Products for African Users*. Her work drives user adoption, retention, and product-market fit across emerging markets.

Lade Tawak is a problem solver and creator working across UX, Product, and AI. Through Limen Labs she provides research, consulting, and coaching. Working globally, she connects people and businesses with the insights and strategic expertise needed to unlock value and achieve goals. With over 10 years of experience, she helps businesses at thresholds and in transition to discover the answers and actions that create success. She is also an international speaker, writer, and community builder.

Hillary Omitogun is a Senior UX Researcher, strategist, and community builder whose work sits at the intersection of human behavior, complex systems, and business decision-making. Professionally, she has worked across financial services, fintech, and consumer technology, leading research on high-stakes journeys and compliance-heavy flows. Hillary is also the founder of *HerSynergy Tribe*, a professional community built to support the growth of mid-senior career women across EMEA.

Part 1

Foundations

Chapter 1

Africa as a Fertile Ground for Innovation

Mama Ngozi stood by her stall in the bustling Balogun market in Lagos, where she sold heaps of fresh yams and peppers. She pulled out her cheap smartphone to open a new app that would let her accept cashless payments. The application icon spun for two whole minutes on the weak connection. A customer grew tired of waiting for the transaction to start. He walked away, and Mama Ngozi felt frustration rise.

The app screen finally appeared, but the interface was difficult to navigate. She fumbled through multiple menus, looking for the "Generate a payment link" button. Her school-age son had shown her the steps the previous week, but the long sequence proved difficult to remember while serving customers. Mama Ngozi put the phone away and moved on to the next customer, accepting cash this time.

She had been told of the power of digital payments, but the system confused her and cost her time. The product failed to adapt to her reality—slow network speeds, limited digital experience, and expensive mobile data. The app was a clever idea, yet it failed to serve the market trader it intended to help.[1]

DOI: 10.1201/9781003648192-2

FIGURE 1.1 The cost of delay: A frustrated Mama Ngozi pleading with her customer to wait while the payment app loads.

The Rise of African Tech

Digital adoption has surged across Africa over the last ten years, fueled by higher mobile phone use and a young, increasingly connected population. Internet penetration in sub-Saharan Africa climbed up from 6% in 2010 to 34% in 2024[2]—a growth driven primarily by mobile broadband (66% of Africans have a mobile device as of 2024[3]). This surge in connectivity has granted millions of people access to information, services, and opportunities previously unimaginable.

The startup ecosystem is also experiencing rapid growth. The number of funded startups rose from 125 in 2015 to 311 in 2019[4]—a 148% jump in four years. This indicates that enterprises are becoming more mature and tackling important challenges. Investment continues to flow into the continent, proving its immense promise: African startups raised $2.2 billion in 2024, despite a 25% dip from the prior year. Although this figure is only 0.6% of global startup funding,[5] it represents substantial value and impact in the ecosystem. We can imagine what more can be done if funding continues to increase.

Africa's technological growth represents immense potential but realizing it requires more than replicating what has worked in other markets. That's why innovators on the continent are pioneering new solutions born of necessity, often taking a fast and aggressive approach to product development.

The "Next Billion Users"

The millions of Africans gaining internet access form part of what the tech industry calls the "next billion users."[6] Beyond signifying untapped markets, this phrase signals a birthplace for innovation where designing for edge cases means designing for all.

In this chapter and the next, we examine case studies that illustrate why we should not simply adapt Western products for African users. Instead, we must recognize that the constraints across much of Africa—scarce infrastructure, low digital literacy, varied languages, cultural contexts, and unique socioeconomic realities—are forcing local innovators to design for resilience, inclusivity, and adaptability. This design philosophy will benefit the entire global tech ecosystem, not just African markets.

The solutions being built in Africa are born from need, so they often perform better than solutions that merely copy and paste what has worked in San Francisco, Berlin, Sydney, or other parts of the world. They show greater efficiency, cost-effectiveness, and real-world usefulness in challenging environments, leading to increased investment interest in this "next billion users" market.

One barrier to adopting this design philosophy is investor pressure. International VCs—a major source of funding for African startups—often prioritize rapid scale and expansion across countries. This is counterintuitive as hyperlocalization—adapting to local payment behaviors, languages, cultural norms, and infrastructure realities in each context—is necessary for adoption in the long run. A product that tries to scale too quickly across diverse markets risks being poorly optimized for all of them, leading to weak adoption, high churn, and wasted resources.

EXPERT INSIGHT: SEYI ONIYITAN ON MODULAR ARCHITECTURE

To scale across countries, companies like RiseVest had to abandon the "rollout" mentality and treat every market based on its local requirements. To comply with regulations and make locally relevant choices, creating a modular compliance architecture works best. Seyi Oniyitan's team at RiseVest planned for regional expansion by migrating to a modular, configurable architecture. Rather than a rigid "one product for Africa," they built a core platform for universal functions such as authentication and data storage with flexible regional layers. This allowed them to switch on specific KYC flows for Kenya or mobile money options for Ghana, ensuring the product complied with local identity regulations and payment preferences from day one.

Innovation in Adversity

African startups do more than cope with constraints—they innovate through them. For example, infrastructure gaps, such as unreliable internet and inconsistent electricity, have led to the creation of sophisticated low-bandwidth applications, offline capabilities, and solutions that bypass traditional infrastructure completely.

African startups do not just cope with these limitations. They INNOVATE through them.

Similarly, they design products to fit diverse or unique user behaviors, such as phone sharing, using power packs, and turning off data to conserve battery and data. This includes creating multiple profiles, enabling offline access, and caching essential user data.

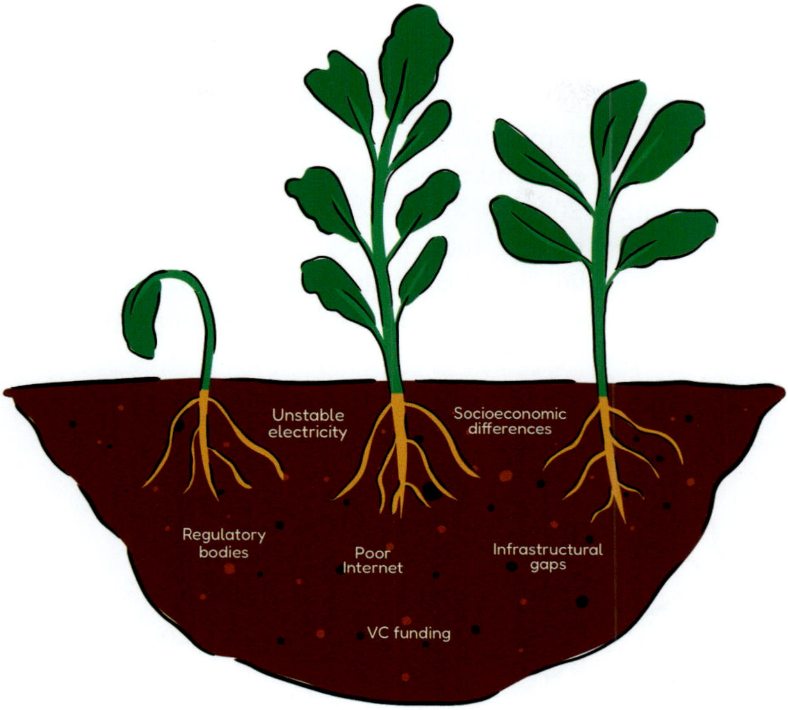

FIGURE 1.2 Africa's tech ecosystem keeps thriving despite its multiple unique challenges.

We will discuss several strategies successful companies employ when building for and with Africa: hyper-local models, community-based agent networks to build trust, workarounds for infrastructure limitations, and design decisions that meet real user needs beyond merely functional apps.

These strategies are grounded in a grassroots approach: products must be designed for and with the everyday person who is not highly digitally literate. This often means resisting the urge to automate everything and instead understanding how processes work traditionally before digitizing a small part.

Moniepoint exemplifies this grassroots philosophy. Rather than requiring agents to complete the onboarding process entirely online, the company chose manual, in-person onboarding. These adaptations drive digital adoption by improving accessibility and usability and fostering participation in the digital economy.

Africa's Big Four Markets: Nigeria, Egypt, South Africa, Kenya

Four markets—Nigeria, Egypt, South Africa, and Kenya—dominate Africa's tech ecosystem, attracting the majority of investment. They are Africa's largest economies and accounted for over 80% of Africa's startup funding in 2025.[7]

Nigeria leads in both total money raised and deal count. It attracted $520 million and 103 equity deals in 2024.[8] Its sheer market size and entrepreneurial spirit create a compelling environment for tech ventures.

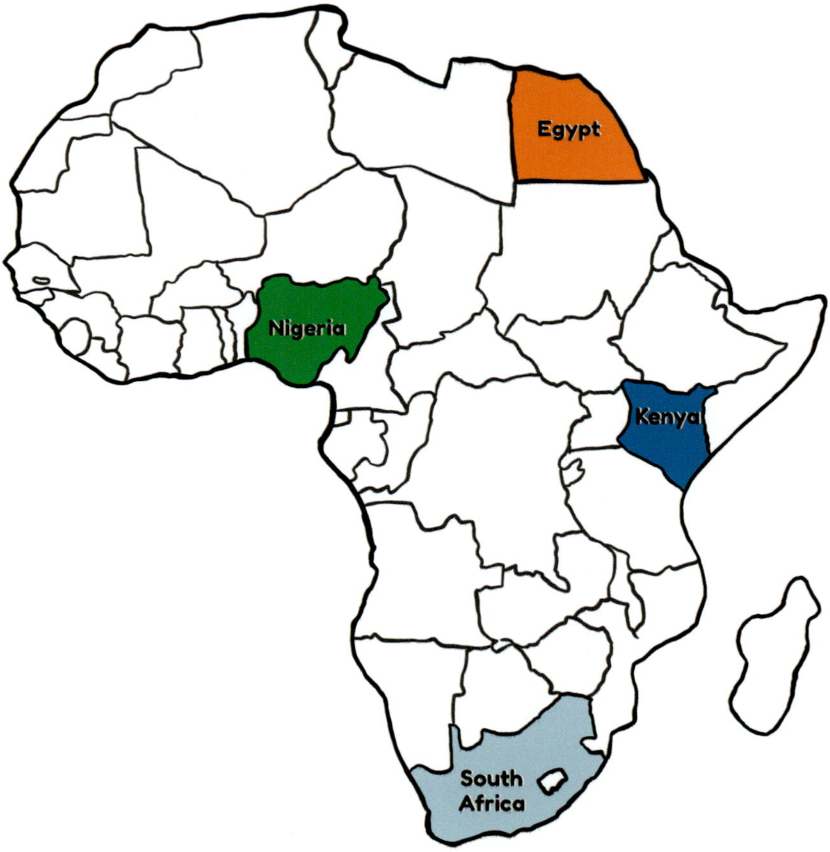

FIGURE 1.3 Africa's big four economic markets.

TABLE 1.1 Africa's "Big Four" Tech Markets[11]

	Nigeria	Kenya	South Africa	Egypt
Population (2024)[12]	~232 Million	~56 Million	~64 Million	~116 Million
GDP (2024)[13]	~$252 Billion	~$120 Billion	~$401 Billion	~$389 Billion
Electricity access (% of population, 2024)[14]	61.2%	76.2%	87.7%	100%
Average cost of 1GB data (2025)[15]	$0.39	$0.59	$1.81	$0.65
Adult literacy rate (% of population)[16]	70% (2024)	82% (2000)	91% (2024)	79% (2022)

Egypt saw the fastest increase in deal activity. From 2023 to 2024, it had a 48% year-on-year increase in equity deals,[9] signaling new strength in its VC ecosystem. Government investment in ICT infrastructure and digital change programmes supports this growth.

Kenya has become a leader in debt funding. It secured 38% of the total debt funding and 31% of the total deal count for Africa in 2024.[10] Its pioneering mobile money system, M-Pesa, has created a culture of digital financial services.

South Africa has more developed infrastructure. Innovation here often happens in sectors like enterprise software and e-commerce, but it still struggles with issues of inclusion and accessibility.

Table 1.1 provides a snapshot of the infrastructural, economic, and sociocultural factors that define the Big Four markets. For product builders, these differences represent the specific *adversity profiles* that the APR Framework™ seeks to address.

Four Trends and Challenges Shaping Africa's Digital Economy

This section previews the key themes explored in Chapters 3 through 6, which examine these challenges and the solutions developed to address them.

Mostly "Mobile-Only" Landscape

Calling Africa "mobile-first" doesn't fully capture the reality. It is more accurately a "mobile-only" continent for most of its connected population.

The mobile phone has become the primary—and often only—gateway to information, commerce, education, and social interaction for millions, as personal computers and fixed-line broadband internet are unaffordable luxuries. Almost 1 in 2 Africans have an active mobile broadband subscription,[17] yet overall internet penetration is around 38% across the continent.[18] This reality reshapes user behavior, necessitating a fundamentally different approach to designing products.

Designers must embrace a "mobile-first" design philosophy where every element is carefully optimized for the constraints of a mobile screen, often a small one. Lightweight applications are also important. Data costs can be too expensive, and device storage is often limited, so heavy apps that use too much data or memory are quickly deleted or abandoned. For product builders, this means a constant focus on efficiency: optimized code, minimal data transfer, and simple user journeys.

For the average African user, their mobile data bundle is a limited resource—every megabyte matters. This means prioritizing text over heavy images or video and aggressively compressing files. Design must use asynchronous data loading to keep the product usable even with poor connectivity. The design must support a wide range of device capabilities, as low-end Android devices and feature phones still make up a large share of the market. This often means keeping basic functionalities simple and accessible.

The reliance on SMS/USSD integrations is a key example of this adaptability. While USSD technology is older than modern digital technology, it remains a foundation of digital services in Africa. Since USSD works without an internet connection, it is especially used for financial transactions, making it important in areas with no internet access. Successful products often offer USSD as a main or secondary channel for core functions. M-Pesa's success in Kenya shows the power of USSD in delivering important services to the masses.

FINANCIAL SERVICES (E.G., M-PESA IN KENYA)

M-Pesa is a social phenomenon that goes beyond a mobile money service. Its success comes from its radical simplicity and profound accessibility. It was built primarily on USSD technology, eliminating the need for smartphones or bank accounts. Its genius was using existing social structures, like the mobile (feature) phone and a vast network of human agents. This created a financial ecosystem that fit seamlessly into daily life.

What is intuitive is highly contextual, as the expert opinions in Chapter 2 illustrate. We often see that many new digital users may not understand concepts like "swiping" or "pinching." As such, designs should adopt more familiar icons, simple tap-based interactions, and linear flows that copy real-world processes. Onboarding must be visual with minimal text, relying on cues rather than excessive instructions.

Mobile phones are often shared among family members and charged in public spaces, so products must include strong security measures and clear logout procedures.

This deep reliance on mobile has created a culture of innovation. This culture prioritizes efficiency, accessibility, and resilience. African designers are forced to solve problems that their counterparts in other markets never face. This makes the African mobile-first landscape a proving ground for truly universal digital products. The services made here are designed for the lowest level of connectivity. They are naturally more robust and inclusive, offering valuable lessons for global product development.

Connectivity and Infrastructure

Despite high mobile phone use, unreliable internet, inconsistent electricity, and unaffordable devices remain big obstacles, as detailed in Table 1.1. For instance, an estimated 45% of Africa's population lives further than 10 km from fiber network infrastructure[19]—a higher percentage than on any other continent.

Research shows that ICT infrastructure increases digital literacy, while social inequality decreases it,[20] underscoring why

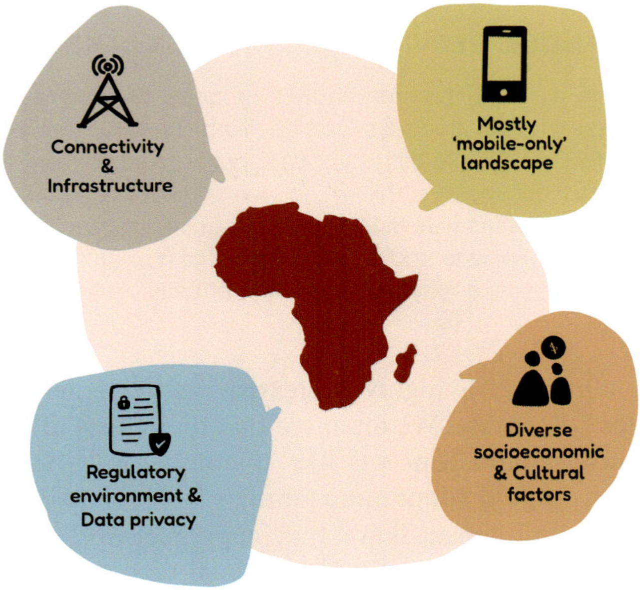

FIGURE 1.4 Trends and challenges shaping Africa's digital economy.

connectivity and infrastructure require the same policy attention currently given to sectors like fintech and healthcare. Policies on digital inclusion can help address availability and affordability barriers related to connectivity and infrastructure.

However, the goal is not to wait for infrastructure to improve before building products, but to build products that work within current realities while infrastructure catches up. This means developing low-bandwidth and offline-first services, as well as adopting designs that use the least amount of data, work well on all connections, and account for power supply problems.

Diverse Socioeconomic and Cultural Factors

The belief that Africa is one uniform entity is a major misconception. Africa is a continent of 54 sovereign nations, each with its own distinct languages, cultures, economic realities, and societal norms. This diversity means that products that try a one-size-fits-all approach across the continent are almost certain to fail—success requires deeply localized, culturally sensitive design strategies.

Digital literacy, for example, varies across regions. Users in highly urban areas may be digitally experienced, while users in rural settings may be using a smartphone for the first time or may never have used one. Accommodating this limited digital literacy would require intuitive, visually driven interfaces with options for voice-guided instructions. These differences in digital literacy between urban and rural areas require a design that can scale across various literacy levels, for instance, using USSD interfaces or avoiding hidden gestures in mobile apps.

Socioeconomic status also varies among users in different countries. Most notably, there is a socioeconomic gap between those who design products and those who use them. Yann Le Beux of YUX Design puts it this way:

> Most of the decision makers sometimes don't realize who they're building products for. Most of them are in the top 1% of the population in their countries, they've been educated abroad, have good salaries, and nice devices, and they build for people they don't understand. This disparity of incomes is a source of the poor experiences.

This disconnect shows that good intentions are insufficient. Builders must actively bridge this empathy gap by prioritizing hyper-local, context-driven research to fully understand the daily realities—data costs, low literacy, and shared-device usage—that define user experience for the majority of Africans.

Payment method preferences also vary across the continent. Although cash is preferred in most countries, mobile money is popular in East Africa (particularly Kenya), while bank transfers are popular in Nigeria. Credit card use, on the other hand, is low across much of the continent. This is due to a lack of trust in online transactions and limited access to formal banking. A platform that does not support cash options or makes the experience difficult faces a huge obstacle.

Jumia, often called the "Amazon of Africa," has adopted varying strategies across different countries. Their earlier success was built on hyper-localized inventory, delivery logistics tailored to each market's unique infrastructure, promotion of local sellers to build trust and community, and diverse payment methods—for example, cash-on-delivery in cash-based ecosystems like Nigeria and mobile money in Kenya. Even a telecommunications giant like MTN, which operates across 18 African markets, constantly changes its service offerings, adapting its data bundles and marketing strategies to fit distinct cultural and economic landscapes.

E-COMMERCE COMPANIES LIKE JUMIA AND KONGA

E-commerce companies like Jumia and Konga have faced significant challenges operating across African markets. However, they have also achieved considerable success through their adaptive design choices. This includes

- Multiple Payment Methods: Extensive cash-on-delivery options alongside digital payments
- Logistics Innovation: Strong delivery networks that address fragmented addresses and unreliable postal systems
- Visual-First UX: Clear product imagery and simplified checkout flows
- Accessible Customer Service: Support channels catering to varying digital literacy levels

The collectivist nature of many African societies affects how products are used. Features that simplify shared accounts, group payments, or community-based services often work well. Trust, typically built on personal relationships, is paramount in this context. Products that use social proof, peer recommendations, or agent networks achieve adoption faster than those relying solely on brand reputation.

Regulatory Environment and Data Privacy

The regulatory environment across Africa is changing, and this affects product design and operation greatly. Policy reforms and digital trade protocols are actively shaping how technology businesses can function. For founders and builders, understanding this framework is fundamental to building trust and sustainable operations.

Efforts to create comprehensive legal frameworks, especially for data privacy and consumer protection, are expanding across the continent. The "Big Four" alongside other countries like Malawi, Zimbabwe, Senegal, and Morocco have their own data protection guidelines similar to other global frameworks such as the EU General Data Protection Regulation (GDPR).[21] Though details vary, most privacy and data protection regulations center on similar principles: user consent requirements, rights to data erasure, and mandatory breach notification.

The adoption of the eight annexes to the African Continental Free Trade Area (AfCFTA) Protocol on Digital Trade in February 2025 promises clearer frameworks for digital trade, which could simplify cross-border operations for tech companies.

Compliance may pose a major challenge due to regulatory fragmentation. While there are efforts to harmonize these regulations at regional levels—such as the African Union Convention on Cybersecurity and Personal Data Protection and the ECOWAS Supplementary Act on Data Protection—rules still vary across the continent, making it difficult to achieve consistency across countries and requiring local legal expertise from companies looking to scale. Companies can navigate this challenge by applying the strictest regulations across all markets to ensure compliance.

Prioritizing compliance is important not only for legal reasons but also for building user trust. Users are more likely to adopt products from companies that handle their personal information responsibly.

This regulatory environment requires careful product design decisions. Product builders must integrate "privacy by design" principles from the start, considering data collection, use, and storage practices.

Beyond data privacy, other regulations influence product design. Payment system regulations affect how payment gateways and fraud prevention measures are designed. Consumer protection laws influence how products communicate with users and handle complaints. Cross-border data flows and trade protocols influence where data centers can be located.

Product teams must work with legal experts to translate these legal requirements into tangible UX flows. Legal language should be broken down into functional requirements, then mapped against the user journey to identify every relevant touchpoint where data is collected or processed. Through iterative, cross-functional design sprints, legal experts (usually an internal compliance or risk team) validate the regulatory requirements, while designers advocate for simplicity and ease of use. This balancing act ensures that high-stakes compliance requirements, like identity verification, are translated into steps that align with the user's digital literacy and cultural context.

Overall, success requires local legal expertise combined with product strategy. This means that product teams must continuously monitor regulatory changes and adapt their products accordingly. As such, compliance must be integrated into the design process from day one.

Africa Needs a New Design Framework!

This combination of factors—a mostly mobile-only audience, unreliable power, high data costs, and diverse cultural needs—exposes the current limits of global design approaches, which assume high digital literacy, constant connectivity, and cultural sameness.

Designers building for Africa—and for a more inclusive global web—must build technology that is simultaneously accessible to digitally inexperienced users yet robust enough to withstand infrastructure failures. This challenge requires a new mental model that begins with adversity as a starting point rather than an edge case. This is why we created the **African Product Resilience (APR) Framework™**. It provides clear, actionable steps for any innovator in the African space to rethink these challenges and turn them into resilient product design.

The APR Framework™: Engineering for Adversity

When Iwalola was a researcher in the image technology department at Transsion, the team in China had just launched a new phone camera with advanced facial recognition. After months of R&D, they had really high hopes for its portrait mode.

FIGURE 1.5 The African Product Resilience (APR) Framework™.

It worked perfectly in test labs under studio lights, so they thought it was ready to launch.

The first field report from Lagos revealed a major flaw—testers complained that the new camera made their photos look "flat" or "ghostly." The algorithm, trained on global datasets, was failing. It overexposed and whitened dark skin tones, smoothing textures that users wanted to keep, and removing the subtle facial details that were culturally important.

The problem was rooted in data bias. The camera was optimized for light skin tones and did not recognize the wide variety of melanin-rich African skin tones. The team had to create a new mandate: retrain the camera's color science using thousands of local portraits taken under harsh sun, low indoor light, and varied outdoor environments. This failure showed that even hardware and image algorithms must be hyper-localized to gain user trust and acceptance.

The African Product Resilience (APR) Framework™ challenges the assumption that products designed for stable Western markets can simply be ported to Africa. Instead, APR™ asserts that products must be **engineered for adversity**. This means building solutions that thrive in conditions marked by low bandwidth, inconsistent power, and diverse digital literacy levels. The APR Framework™ helps us identify innovative solutions that create universal usability, as Africa is the proving ground for truly adaptable and user-centric technology.

The framework defines four pillars for actionable design (Table 1.2).

TABLE 1.2 The Pillars of the African Product Resilience Framework™

Pillar	Element	Focus
Building for Digital Access	Mobile-first optimization	Prioritizing efficiency and minimal data usage, as most users face challenges with lower-end devices and expensive mobile data
	Infrastructure agnosticism	Building resilient solutions that work across all connectivity states, including limited or no internet. Involves adopting an "offline first" mentality
Building for Trust	Trust-centered interactions	Building confidence in digital products through human validation, regulatory compliance, and privacy by design
Building for Digital Literacy	Literacy scalability	Accommodating users with varying levels of textual and digital literacy and varying levels of experience with digital tools through voice, illustrations, videos, and content in local languages
Building for Sociocultural Context	Hyper-local customization	Adapting to local culture, customs, and behaviors in text, imagery, and design patterns, because what works in one context may not resonate in another

Building for Digital Access[22]

Mobile-First Optimization

Digital access addresses the mobile-first reality of most Africans. The vast majority of digital interactions in Africa happen on a mobile phone, which is often the user's only computer. This demands a mobile-optimized design approach.

This pillar dictates treating mobile as the primary, high-priority channel, not an afterthought. Design must be ruthlessly efficient to accommodate the constraints of low-end smartphones, where limited storage and expensive data are major factors for app adoption and retention.

Designers must prioritize data efficiency and simplicity. Every feature must work reliably on smaller screens, older and less powerful devices, and older operating systems. People without smartphones should also be able to access basic functionalities.

EXPERT INSIGHT: TOYIB MUSTAPHA ON BUILDING FOR LOW CONNECTIVITY

For resource-intensive features, such as real-time location sharing, product teams must manage background processes. To prevent low-end devices from lagging, overheating, or crashing due to insufficient RAM, the app must ensure background services don't run constantly when the app is not in the foreground. This includes:

- Reducing Accuracy: Prioritizing performance by setting the GPS accuracy level to medium
- Time-Boxing Real-Time Data: Restricting real-time location updates to a specific time window (e.g., 10 minutes while actively using the app), then holding on to the last known data
- User-Triggered Fetching: Only triggering real-time data or API fetching when the user performs an action like "drag to refresh" or re-enters the app

This approach ensures that the app minimizes burden on the device and conserves battery and data, thereby improving retention rates among users on feature-limited hardware.

Example: An onboarding process should use minimal, compressed imagery. It should also have an immediate fail-safe for low storage, alerting the user before installation.
Question to Ask: Does this feature add enough value to justify the cost in data consumption and device memory?

Infrastructure Agnosticism

Digital access also includes larger infrastructure and connectivity challenges.

Connectivity is unpredictable, so products must work even when the connection drops entirely. Infrastructure agnosticism means the product should function equally well across different technical environments, including 2G/3G networks. This requires building offline capabilities for core functions and using channels like USSD and SMS for critical transactions when data is unavailable.

Example: A healthtech app allows community health workers to record patient data offline and automatically syncs the data when the phone reconnects to a reliable network.
Question to Ask: Can the user complete the single most important task even if the phone has zero data and one bar of 2G or 3G signal?

Building for Trust[23]

Digital services require a high level of trust from African users. This trust is often built through human validation (e.g., community and personal validation), not just digital security badges.

This pillar emphasizes integrating community and human elements directly into the digital experience to build user confidence and drive adoption. This includes verifiable agent networks (particularly for widespread consumer products in financial services, healthtech, and other high-stakes sectors) or in-app features that enhance personal security.

Example: A payments app uses a physical agent network for cash deposits. The app displays the agent's verified badge and transaction history to reassure users.

Question to Ask: How can we replace the handshake of a physical transaction with a digital confirmation that uses a trusted human element?

EXPERT INSIGHT: TOCHY EMEREOLE ON COMMUNITY-DRIVEN TRUST

Trust is one of the biggest barriers to digital adoption in Africa. There's a saying in marketing: *people don't follow logos; they follow people.* Rather than relying on faceless ads, I use a community-led marketing strategy, focusing on micro-influencers and community leaders as drivers of trust, since they already hold credibility within their circles. We facilitate the transfer of trust from established community relationships to our digital platform.

Our approach:

1. Structured Onboarding: We thoroughly educate these advocates on our vision, mission, core values, and the specific problems we solve for their communities
2. Cultural Translation Rights: We give them creative latitude to adapt our messaging, imagery, and communication style into language and formats that resonate authentically with their audiences—they know their communities better than we do
3. Collaborative Control: They have meaningful input (including veto power) on campaign elements, creating a balance between brand consistency and their unique voices

This strategy has consistently delivered outstanding results. For instance, users acquired through these community-led campaigns tend to show higher trust and product stickiness; many begin using the product immediately and conduct large transactions from day one, without the cautious "testing" behavior seen in paid ad segments. Data from multiple fintech and e-commerce campaigns showed that while paid-ad users typically reach stickiness by their fifth usage, community-acquired users reach it as early as the third interaction.

Over time, I've also found that micro-influencers drive conversions, while macro-influencers excel at brand awareness. Leveraging both in the right balance creates a holistic trust-centric ecosystem. By letting trusted community voices carry our message, we build belonging, credibility, and emotional assurance, the true foundations of trust in digital services.

Building for Digital Literacy[24]

Digital literacy varies widely, even within one city. Literacy scalability requires interfaces that accommodate users who may be interacting with a smartphone for the first time. The design must be intuitive with simple navigation, clear and well-tested iconography, and linear flows that mirror real-world steps, reducing cognitive load for new users. It should minimize reliance on complex vocabulary or abstract digital concepts.

Adopting this pillar means that interfaces must be immediately understandable to users with varying levels of digital and textual literacy.

EXPERT INSIGHT: YANN LE BEUX ON BRIDGING THE DIGITAL LITERACY GAP

Adding the right "helpers" can greatly improve the user experience. A mix of English text, good iconography, and an audio helper creates a lighter, simpler experience than trying to build a fully AI-driven interface. This is especially true because current NLP tools don't yet work well for many African languages. By combining these three elements—text, image, and voice—builders can create a "multimodal" safety net that meets the user where they are. An example of this is Wave in Senegal: their app includes manually recorded audio explanations in local languages for special offers.

Example: A loan application uses voice guidance in a local language alongside clear, pictorial icons instead of traditional menu labels.

Question to Ask: If a user only saw the icons and heard the voice instructions, could they complete the necessary action without reading any text?

Icon-based, simplified UIs significantly improve task success rates and reduce completion times for users with lower digital literacy, making products more inclusive and effective.

Building for Sociocultural Context[25]

Africa is not one market—it's 54 countries with distinct cultures, languages, and social practices. Design cannot treat this diversity as a minor variable to be addressed through language translation. Instead, it should aim for true localization, which involves adapting the product's core functionality to local behaviors, economic realities, and payment preferences. This includes supporting unique payment types, such as cash-on-delivery, and allowing social practices like haggling to be part of the user journey.

EXPERT INSIGHT: KOME SIDESO ON THE MYTH OF "FREE IS BETTER"

> In some markets, if it's free, people won't show up. They don't believe in it.

A common assumption in product growth is that "free" is the ultimate motivator for adoption. However, pricing is often a deeper sociocultural signal of value and legitimacy that can vary across markets. While a "free" offer in Nigeria might result in a packed event or rapid app downloads, the same strategy might fail spectacularly in Kigali, Rwanda. In the Rwandan context, we observed that users often met free services with suspicion. Payment acts as a verification of quality and intent, and free products might be perceived as scams.

When entering a new market, don't assume that lowering the price to zero will lower the barrier to entry. In high-trust or highly structured societies, a subsidized fee is often more effective than a free offering because it respects the user's need for a "legitimate" exchange. Builders must design their economic model to match the socioeconomic class and cultural expectations of the local environment.

Example: An e-commerce platform allows users to pay via a mix of mobile money and cash-on-delivery, reflecting local payment behavior. It also adapts product recommendations based on regional festivals and events.

Question to Ask: What local market behavior does our product currently prevent or ignore? How can we incorporate that behavior into the digital flow?

The APR Framework™ provides a clear roadmap to creating products that are truly useful and trusted by African users.

Successful products reflect these framework elements, embracing constraints and transforming into competitive advantages. Hyperlocal, user-centric designs offer lessons for global builders, making products more resilient, usable, and accessible for everyone. This is the foundation upon which this book is built.

Key Takeaways

1. Africa is proof that systemic problems can be catalysts for innovation, enabling and not hindering resilient technology development
2. For many users, "mobile-first" essentially means "mobile-only"
3. In low-trust environments, digital adoption is often driven by human intermediaries who bridge the gap between digital platforms and users
4. To succeed, teams should adopt the African Product Resilience (APR)™ Framework, which is built on four pillars: Building for Digital Access, Building for Trust, Building for Digital Literacy, and Building for Sociocultural Context

Notes

1. Note: This is a fictionalized story based on common user feedback gathered during real-world usability tests.

2. "Individuals Using the Internet (% of Population)—Sub-Saharan Africa," World Bank Open Data. Accessed May 28, 2025. https://data.worldbank.org/indicator/IT.NET.USER.ZS?locations=ZG
3. "Mobile Phone Ownership," ITU, November 10, 2024. https://www.itu.int/itu-d/reports/statistics/2024/11/10/ff24-mobile-phone-ownership/
4. Fredrick Eghosa, "Trends of Funding to African Startups (2000–2024): Investment & Growth Insights," *Intelpoint* (blog), May 28, 2025. https://intelpoint.co/blogs/trends-of-funding-to-african-startups/
5. Fredrick Eghosa, "Trends of Funding to African Startups (2000–2024): Investment & Growth Insights," *Intelpoint* (blog), May 28, 2025. https://intelpoint.co/blogs/trends-of-funding-to-african-startups/z
6. Peeyush Ranjan, "An Anthology of Insights, for a More Inclusive Internet, *The Keyword* (Google blog), March 15, 2022. https://blog.google/technology/next-billion-users/anthology-insights-more-inclusive-internet/
7. Max Cuvellier Giacomelli, "2025 IN REVIEW - Mapping the Money," *Africa: The Big Deal* (Substack), January 13, 2026. https://thebigdeal.substack.com/p/2025ir2

8. Partech, "2024 Partech Africa Tech VC Report: With US$3.2B Raised, African Startups Show Resilience Despite 7% Drop in Funding," Accessed July 31, 2025. https://partechpartners.com/news/2024-partech-africa-tech-vc-report-with-us32b-raised-african-startups-show-resilience-despite-7-drop-in-funding/
9. Partech, "2024 Partech Africa Tech VC Report: With US$3.2B Raised, African Startups Show Resilience Despite 7% Drop in Funding," Accessed July 31, 2025. https://partechpartners.com/news/2024-partech-africa-tech-vc-report-withus32b-raised-african-startups-show-resilience-despite-7-drop-in-funding/
10. Partech, "2024 Partech Africa Tech VC Report: With US$3.2B Raised, African Startups Show Resilience Despite 7% Drop in Funding," Accessed July 31, 2025. https://partechpartners.com/news/2024-partech-africa-tech-vc-report-withus32b-raised-african-startups-show-resilience-despite-7-drop-in-funding/
11. All data sourced from the World Bank and Best Broadband Deals reflects the most recently available figures at the time of writing.
12. "Population Total: 2024", World Bank Open Data. Accessed February 18, 2026. https://data.worldbank.org/indicator/SP.POP.TOTL
13. "GDP (Current US$): 2024," World Bank Open Data. Accessed February 18, 2026. https://data.worldbank.org/indicator/NY.GDP.MKTP.CD
14. "Access to Electricity: 2023," World Bank Open Data. Accessed February 18, 2026. https://data.worldbank.org/indicator/EG.ELC.ACCS.ZS
15. "The Cost of 1GB of Mobile Data in 237 Countries," Best Broadband Deals. Accessed August 30, 2025. https://bestbroadbanddeals.co.uk/mobiles/worldwide-data-pricing/
16. "Literacy Rate, Adult Total (% of People Ages 15 and Above): 2025," World Bank Open Data. Accessed February 18, 2026. https://data.worldbank.org/indicator/SE.ADT.LITR.ZS.
17. "Facts and Figures 2024: Subscriptions," ITU. November 10, 2024. https://www.itu.int/itu-d/reports/statistics/2024/11/10/ff24-subscriptions/
18. "Facts and Figures 2024: Internet Use," ITU. November 10, 2024. https://www.itu.int/itu-d/reports/statistics/2024/11/10/ff24-internet-use/
19. Roku Fukui et al., "Africa's Connectivity Gap: Can a Map Tell the Story?," *World Bank Blogs*. November 7, 2019. https://blogs.worldbank.org/en/digital-development/africas-connectivity-gap-can-map-tell-story
20. Meiqi Sun, Jiaqi Liu, and Jia Lu, "Digital Literacy in Africa: Exploring Its Relationship With Infrastructure, Policy, and Social Inequality," *African Journalism Studies* 44, no. 3 (2023): 204–25. https://doi.org/10.1080/23743670.2024.2329705
21. We explore privacy, ethics, and regulations in greater detail in Chapter 6.
22. Covered in Chapters 3 and 4.
23. Covered in Chapter 6.
24. Covered in Chapters 3 and 5.
25. Covered in Chapter 5.

Bibliography

Best Broadband Deals. "The Cost of 1GB of Mobile Data in 237 Countries." Accessed August 30, 2025. https://bestbroadbanddeals.co.uk/mobiles/worldwide-data-pricing

Eghosa, Fredrick. "Trends of Funding to African Startups (2000–2024): Investment & Growth Insights." *Intelpoint* (blog). May 28, 2025. https://intelpoint.co/blogs/trends-of-funding-to-african-startups/

Fukui, Roku, Christopher James Arderne, and Tim Kelly. "Africa's Connectivity Gap: Can a Map Tell the Story?" *World Bank Blogs*. November 7, 2019. https://blogs.worldbank.org/en/digital-development/africas-connectivity-gap-can-map-tell-story

Giacomelli, Max Cuvellier. "2025 In Review - Mapping the Money." *Africa: The Big Deal* (Substack). January 13, 2026. https://thebigdeal.substack.com/p/2025ir2

International Telecommunication Union (ITU). *Measuring Digital Development: Facts and Figures 2024*. November 10, 2024. https://www.itu.int/itu-d/reports/statistics/facts-figures-2024/index/

Partech. "2024 Partech Africa Tech VC Report: With US$3.2B Raised, African Startups Show Resilience Despite 7% Drop in Funding." Accessed July 31, 2025. https://partechpartners.com/news/2024-partech-africa-tech-vc-report-with-us32b-raised-african-startups-show-resilience-despite-7-drop-in-funding/

Ranjan, Peeyush. "An Anthology of Insights, for a More Inclusive Internet." *The Keyword* (Google blog). March 15, 2022. https://blog.google/technology/next-billion-users/anthology-insights-more-inclusive-internet/

Sun, Meiqi, Jiaqi Liu, and Jia Lu. "Digital Literacy in Africa: Exploring Its Relationship With Infrastructure, Policy, and Social Inequality." *African Journalism Studies* 44, no. 3 (2023): 204–25. https://doi.org/10.1080/23743670.2024.2329705

World Bank Open Data. "Access to Electricity: 2023." Accessed February 18, 2026. https://data.worldbank.org/indicator/EG.ELC.ACCS.ZS

World Bank Open Data. "GDP (Current US$): 2024." Accessed February 18, 2026. https://data.worldbank.org/indicator/NY.GDP.MKTP.CD

World Bank Open Data. "Individuals using the Internet (% of population) - Sub-Saharan Africa." Accessed May 28, 2025. https://data.worldbank.org/indicator/IT.NET.USER.ZS?locations=ZG

World Bank Open Data. "Literacy Rate, Adult Total (% of People Ages 15 and Above): 2025." Accessed February 18, 2026. https://data.worldbank.org/indicator/SE.ADT.LITR.ZS

World Bank Open Data. "Population Total: 2024." Accessed February 18, 2026. https://data.worldbank.org/indicator/SP.POP.TOTL

Chapter 2

A Primer on the African Context

Kofi runs a busy barbershop just outside Kumasi. Every morning, he buys a small data bundle to manage appointments, receive digital payments, and perform other tasks on his basic Android phone. The network signal near his shop is often poor.

One morning, he opens a new stock-ordering app to check the delivery time for his new clippers, but the screen struggles to load. When it eventually loads, a large, high-resolution advertisement banner for hair products appears first, its vibrant image consuming the last of his data. The app then failed completely with a simple error message: "Connection Lost." Kofi refreshes the page many times. Each attempt cost him time but failed to load the delivery status. The heavy graphics made that simple task impossible.

Kofi sighs and puts the phone away. He has no idea when his clippers will arrive, so he can't plan ahead. The product was technically advanced, but its failure to account for low-bandwidth conditions cost Kofi an hour of work and made him frustrated. The app was built on the assumption that every user has unlimited broadband. For Kofi, every single megabyte spent is money taken directly from his profit.[1]

DOI: 10.1201/9781003648192-3

FIGURE 2.1 The cost of a colorful ad: An upset Kofi is unsure about the delivery of his clippers.

An Overview of the Socioeconomic Landscape and Its Digital Implications

Situational factors determine whether people can use a digital product at all. These factors include access to electricity, network reliability, and the cost of mobile data. Although often treated as peripheral concerns, these factors determine whether a product succeeds or fails.

A product that is "perfectly designed" will still fail if the target user cannot afford to download it or their phone battery dies every three hours. To build products that are usable for Africans, we must treat these external problems as core design constraints.

In this chapter, we will introduce the factors that shape how African users adopt and use digital products. These factors are discussed in greater detail in the following chapters.

Demographics

Africa has the world's youngest population, with a median age of about **19 years.**[2] This demographic presents a huge opportunity for digital products, as young people adopt new technology more readily and form the core base for social media, fintech, and edtech platforms.

This reality should influence how products are designed. Visual aesthetic and interaction patterns must feel modern and culturally relevant to this young group.

The Urban-Rural Disparity

African countries typically have high population growth rates, which means the number of potential users increases each year. However, this growth is unevenly distributed between dense urban centers and vast rural areas.

Urban areas, such as Lagos, Cairo, or Nairobi, show high levels of mobile penetration due to a greater concentration of wealth. Users here often use high-end smartphones and have better access to 4G networks. They expect faster loading times and more complex features. Their product needs might mirror those in Western cities: convenience, on-demand delivery, and complex financial tools.

Users in rural areas, on the other hand, often have different needs. They often rely on feature phones or older, lower-end smartphones. Access to consistent electricity is a daily challenge, and Internet access is often limited or unpredictable. The cost of data is also a big barrier. Consequently, products that succeed here must be extremely data-light and offer core functionality through simple channels like USSD.

Economic Diversity

Economic diversity is another crucial factor. Nigeria alone has millions of people across a wide economic spectrum, with wealth concentrated at the top. This means a large part of the population operates in the informal economy, conducting daily transactions with cash, as they may lack access to formal bank accounts. Products aiming for financial inclusion must accommodate this reality by enabling cash transactions where possible alongside digital payment methods.

Income levels and earning power directly affect digital adoption. For many Africans, the prices of smartphones and data costs are major financial hurdles. A standard entry-level smartphone costs more than many people earn in a month, forcing many to use older devices with limited processing power, outdated operating software, and small storage capacity.

Low income creates two main problems for product designers. First, the user cannot afford expensive hardware. This means the app must be compatible with older versions of Android and work on devices with slow processors and small screens.

Second, the user cannot afford high data costs, making every megabyte the app uses a significant concern. Successful products treat the user's data bundle as a precious resource. Designers must compress images aggressively, limit video autoplay, and remove unnecessary animations. Additionally, they must ask the user for permission before downloading large files. Respecting the user's budget in this way builds trust and reduces the likelihood of the app being uninstalled to conserve storage space or data.

Income and access to infrastructure are inextricably linked. Users with lower incomes often live in areas with poor infrastructure, experiencing less reliable electricity, and more frequent network downtime. Digital products and services must handle these constant disruptions by, for example, saving progress locally and syncing data quietly when the connection returns. The app must never leave the user hanging during a power cut; it must always feel dependable, regardless of external conditions.

Feature Phones Remain Highly Relevant

In Africa, the feature phone is far from dead. It remains a key tool for communication and financial access across many

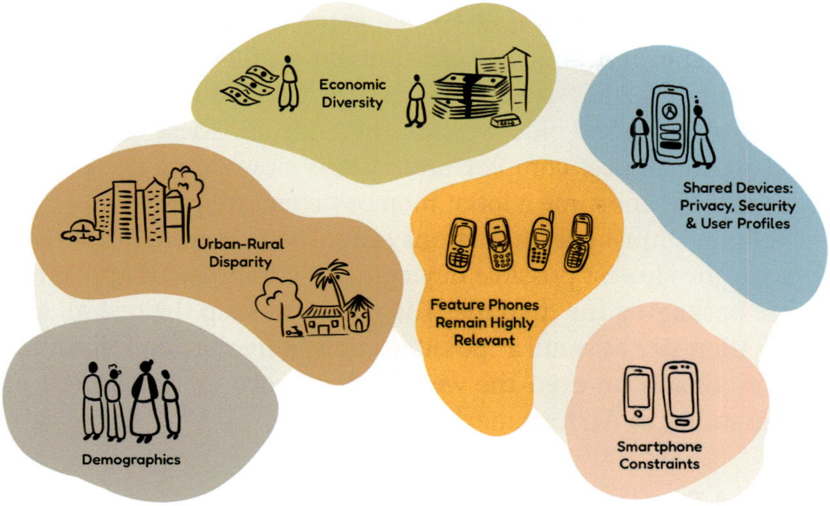

FIGURE 2.2 Africa's socioeconomic landscape.

sub-Saharan countries because of its low cost, long battery life, and durability. Moreover, it only needs a mobile network connection to function.

Feature phones cannot run complex apps, so users access services via Unstructured Supplementary Service Data (USSD). USSD is a GSM network protocol that enables users to send and receive short messages through simple text menus. USSD is not limited to feature phones but also works on any mobile phone, even without an internet connection.

USSD is the foundational UX layer for mobile money services like M-Pesa, powering its early success. Due to the prevalence of feature phones in Africa, product builders need to treat USSD as a primary design channel, not a backup. The USSD journey must be clean and simple, requiring no more than a few steps to complete critical tasks such as sending money. If USSD is treated as the foundation, it can be implemented for core flows. This means designing these core flows using only text menus first, before building smartphone apps. This embeds clarity and simplicity into the service, making the digital product accessible to everyone, regardless of their device type or location.

Smartphone Constraints

Even the cheapest smartphones present challenges. They often have limited RAM—sometimes as low as 1GB or 2GB—small internal storage, and older versions of Android. These constraints affect performance: complex animations slow down the device, and apps that consume a lot of memory crash frequently.

Designers must use animations only when necessary, test their apps on the slowest possible devices, and compress all app assets before packaging the final file. The goal is to keep the overall app installation size as small as possible and ensure the app feels quick and responsive on even the weakest hardware—a crucial test of true usability in African markets.

Shared Devices: Privacy, Security, and User Profiles

The single-user model does not reflect the lived reality in many African communities. Device sharing is common in many African communities because of the high cost of hardware. Families may share one smartphone for education, financial services, and communication, while small businesses share devices among staff members to process payments or access other work tools. Consequently, product designers must not assume that the person holding the phone is the only account holder or the original user, making privacy and security critical design considerations.

Privacy is the first casualty. If an app auto-logs a user in, anyone who picks up the phone has access to personal data, such as financial data, messages, and contact details. Designers should introduce stronger lock mechanisms into their application interface, as a simple device PIN might not be sufficient when sharing is the norm.

Security poses another problem. If a child uses the shared phone to play games, they might accidentally, or even intentionally, delete an important application, send unauthorized messages, or attempt purchases. Products must offer clear user profiles or guest modes to separate sensitive data from everyday functions.

Africa is largely an Android-dominant market due to the high cost of Apple devices. While iPhones are priced beyond the reach of the vast majority of the population, Android phones are available at every price point, making Android the default choice for millions of first-time smartphone owners.

iOS adoption is growing, but it remains a premium product with market share concentrated among upper-income groups and major urban centers. In some countries, the ratio of Android to iOS users can be as high as 9:1.[3] Consequently, products targeting broad population access must prioritize Android optimization. Building for iOS still matters, but Android is the platform for scale and inclusion. Designers must consider this disparity when allocating time and resources for development and testing.

Considerations to Ensure Cultural and Socioeconomic Localization

The product audit checklist helps product teams to examine and question assumptions, ensuring their products or prototypes are well adapted to the cultural and social needs of their target market. It highlights common factors that influence adoption and usability for African users. In the chapters that follow, we will break down each element and guide you on working with them.

PRODUCT AUDIT CHECKLIST

Before launching a product in an African market, use this checklist as a starting point. It is not exhaustive but can be used as a baseline framework to brainstorm additional questions specific to your product and context with your team.

- Literacy: Do key flows like onboarding rely primarily on visual cues and clear icons rather than requiring the user to read dense text?
- Language: Have all core user interface elements been reviewed and confirmed in the key local languages (e.g., French, Swahili, Yoruba) where required?
- Infrastructure and Data Efficiency Audit: What is the total app installation or update size? Small size saves storage space on low-end phones and reduces the download cost.

- Addresses: Does the driver/delivery interface support GPS coordinates and a robust in-app chat for clarifying imprecise addresses?
- Trust: Does the product use local social proof or community-based features to build trust? If the product relies on a physical interaction (e.g., a delivery drop-off, a health consultation), does the interface verify the identity of the person involved (e.g., rider ID, doctor photo) to build trust?
- Privacy: Does the product adhere to the specific data privacy laws of the target country, such as the NDPA in Nigeria or POPIA in South Africa?
- Shared Use: Does the design account for shared devices or communal use with strong security or logout features?
- Payment Adaptation: For services that charge money, are dominant local payment methods supported—**cash-on-delivery** (e-commerce), **mobile money** (M-Pesa, etc.)—not just bank cards?
- Offline Functionality: Does the product provide an offline mode or USSD fallback for critical tasks? This ensures the service is accessible without the internet (e.g., viewing a previous diagnosis, checking class schedule)
- Basic Interaction: To ensure usability in worst-case scenarios, have we tested the core user flow on a low-end Android device with poor signal?

When to Use the List

These considerations are most valuable during the final stages of the design and quality assurance (QA) process.

1. During Design Review: Use the list to review wireframes and mockups. This ensures the visual design addresses the local context before the developers start work
2. For QA Testing: Pass this list to the QA team in the target market. They must confirm that features such as cash acceptance and security measures for shared devices work correctly
3. Before Sign-Off: Use the list before the final launch. This serves as the mandatory sign-off to confirm that all cultural and infrastructure needs have been met

Limitations

This list does not guarantee product-market fit. Success still depends on your product's core value proposition. It has one main limit: it is not a research plan or a replacement for talking to users. It provides the questions to confirm, not the data to discover. You must still conduct fieldwork to understand local behavior.

Case Studies in Adaptation

Successful African startups share a common approach: they deeply research local behaviors and technical limits and build their products to fit the market rather than force the market to adapt. These examples show the APR Framework™ in action—by emphasizing that builders must adopt the pillars of the framework in order to build solutions that fit the market and not the other way around.

Fintech Apps

Moniepoint, OPay, and PalmPay have rapidly grown into super-apps in Nigeria by mastering the bridge between digital and cash economies. Rather than create banking apps alone, they created agent networks—human touchpoints where users can deposit and withdraw physical cash. This directly solves the problem of low bank account penetration. Two design choices distinguish these products:

Agent Network Integration

The agent network is a key UX feature. Their app designs support the agent experience with simple, fast, and transactionally secure flows. Furthermore, the physical experience of giving cash to a trusted agent is digitized inside these apps, which builds user confidence and makes the digital wallet feel safe. Also, these apps are built to work perfectly on cheaper devices, even with a poor connection.

Local Languages and Visual Design

These apps successfully integrate local elements to build familiarity between the product and the users' real world. The visual design within their interfaces and marketing materials uses colours, imagery, and text that resonate with Nigerian culture. Additionally, they use clear, familiar icons for functions like sending money or paying bills. Transaction flows are also linear to reduce the risk of user error. Overall, the design speaks directly to users whose primary mental model may still be based on a physical, handwritten ledger.

Designing for Healthtech

ThinkPlace and the Ministry of Health in Ethiopia faced a challenge regarding data culture: while health workers had extensive data, they felt disconnected from the task of reporting it, resulting in inaccurate or incomplete information. Rather than relying on the existing open-source DHIS2 platform alone, they created a bridge to frontline workers in remote regions to improve data quality and utilization. They solved this problem by creating the Eyta chatbot, a simple digital tool that fetches reported data and makes it accessible.

Conversational Platform Integration

The choice of platform is a key UX feature. Insights showed workers were less likely to engage with complex dashboards or technical language. Instead of a standalone mobile app or micro-website, ThinkPlace developed the Eyta chatbot on Telegram and WhatsApp. This design mimics social media principles to create an interactive, personable experience on channels where users are already comfortable, avoiding the need for complex new technical interfaces.

Visual Simplification and Personalization

The tool prioritizes intuitive graphs, pie charts, and simple text over health informatics technical language to make the data understandable. Eyta transforms quarterly and monthly indicators into familiar visual formats, serving as an accessible extension of the DHIS2 system. Additionally, the tool allows health workers to select and track the indicators they are personally interested in—not just those required by decision-makers—ensuring the data feels relevant and valuable to their specific context.

E-Commerce

E-commerce platforms like Kilimall in Kenya face unique problems. They must build trust with users who are skeptical about buying items they cannot see or touch while managing complex logistics and uncertainties around online payments. They solved this problem by prioritizing:

Product Presentation and Authenticity

These platforms prioritize clear, high-quality product imagery and prominently display authenticity guarantees on the product page, as users need reassurance that the product they see online is the one they will get.

Payment Flows and Customer Support

Payment flows must accommodate multiple payment methods. For example, Kilimall supports card payments, mobile money platforms like M-Pesa, and cash-on-delivery (which remains essential in African markets). Regardless of the method, the payment process must be easy to complete.

Additionally, customer support must be easily accessible through popular local communication channels, such as WhatsApp or dedicated local chat features. East African e-commerce designers recognize that human interaction is necessary for resolving delivery and product quality concerns. The digital process is designed to have a human fallback option.

Problem Identification

In all the case studies highlighted thus far, the teams had to clearly identify the problems they wanted to solve before moving on to the how. We have found that this step—identifying the user problem—is often a critical roadblock for product teams.

When Iwalola first started designing a simple banking app for micro-merchants, the team assumed the biggest problem was usability. They spent weeks simplifying the dashboard, working on making the icons clearer and the visualizations smoother, and thought the product was beautiful. But after launching the pilot, adoption was slow. They didn't understand the failure until they visited the market and watched a merchant try to log in. The login screen failed repeatedly. The developer blamed a slow backend until the merchant showed her phone. She was running the app on 12MB of data, which she planned to extend for the whole week. Meanwhile, the app was performing an automatic asset refresh in the background, silently consuming her entire data plan.

The problem was one many teams make regularly: confusing usability (is the app easy to use?) with accessibility[4] (can the user afford to use it?). This shows that testing our assumptions with the right questions *before* designing and writing code is essential. The five-question guide helps ensure you don't make that foundational mistake.

Designers seeking to innovate in adversity must deeply understand the user and their real world. These questions will help guide designers and innovators through identifying the real pain points:

Defining the UX Problem: Five Questions to Guide Problem Identification

1. Is the user struggling with the product's design (usability), or with accessibility (data cost, device quality, literacy)?
2. How does the user solve this problem today without our product, for example, using an agent, writing it down, or using cash?
3. If this is the user's only computer (their mobile phone), what critical, non-app functions must we design around, for example, camera use for verification?
4. What specific feature in our product relies on trust, for example, making a digital payment or providing personal data? How can we use local community structures or personal interactions (agents) to solidify that trust?
5. How much data (in MB) and how many steps does it take a user to complete the most critical task? Is this cost acceptable given the local pricing of data bundles?

When to Use This Guide

This checklist is most valuable during the discovery and definition phases of product development.

1. At Project Kickoff: Use it at the start of a new project to establish the right research priorities and prevent the team from defaulting to Western assumptions
2. During User Research Synthesis: Use it to frame your findings. The questions will help your team to move beyond surface observations to find the root cause of user difficulties, especially in areas of low connectivity or literacy
3. For Feature Prioritization: When choosing which features to build, use the "Cost of Action" and "Usability versus Accessibility" questions to prioritize resilience over novelty

How to Use the Questions

The questions are designed for group discussion and active research. **They are not a survey.** They are prompts for deeper thought and investigation.

- Usability versus Accessibility: This is the first critical filter. If your team finds that users cannot access the app because of data costs, no amount of UI improvements will help. The solution lies in Infrastructure Agnosticism (e.g., USSD integration), not Literacy Scalability (icon changes)
- Alternative Behaviors: Ask researchers to document the analog process. For instance, how do market women currently save money without your app? What steps do they take when they pay with cash? Your digital product should be designed to support and streamline that familiar flow, not replace it. You're looking for the existing mental model to simplify the digital one
- "Only Computer" Reality: Consider device limitations when making product and design decisions. Don't assume the user has a desktop for complex tasks. If the mobile camera is the only way to verify identity, your app's design must make that camera process simple and data-light. Test camera functionality under poor lighting conditions

- Trust Barrier: Find the biggest friction point where users might hesitate to transfer money or provide their ID. If your product relies on trust in a digital interface, reinforce that trust with physical, local elements. This might be a verified agent's photo or a message from a local community leader
- Cost of Action: This is a metric that developers and designers must jointly measure. Calculate the average MB used for critical tasks (e.g., sending money). This must be balanced against local data bundle costs. Also, keep steps as few as possible as with every additional tap or page load, the risk of network failure interrupting the experience goes up

Limitations of This Guide

This guide has two main limitations.

1. It requires context. These questions are useless without active research. They must be answered with real, verifiable data from interviews and field studies in the target country. Do not answer them based on office assumptions
2. It doesn't replace technical planning. These questions set the design requirements—they don't tell the developer which specific technology to use. The developers must still plan the technical implementation for offline syncing and data compression

Key Takeaways

1. The African user's socioeconomic and cultural context is often the key constraint to the adoption and usage of digital products
2. Product failure is often rooted in accessibility constraints based on factors like high data cost, poor network, and device type, rather than just interface design
3. External factors—such as unreliable electricity, expensive data, and poor network coverage—are core design constraints and not simply edge cases
4. Successful products adapt to the needs of local realities

What Next?

Action Checklist

- Audit the existing analogue process (if any) to document exactly how users solve that problem today using cash, paper, agents, or physical actions
- Consider designing solutions to validate, simplify, and elevate that familiar behavior/mental model, rather than jumping to replace it entirely
- Use the Problem Identification Guide to filter your research findings
- Test your product for the Cost of Action—calculate the exact amount of data (in MB) and the number of taps required for the user's most critical task
- Treat the user's data bundle as a valuable resource and prioritize reducing this cost aggressively
- Use the Cultural & Socioeconomic Localization Checklist to audit your solution

Reflection Questions

- Is the primary struggle our users face usability or accessibility?
- How does our user currently solve their problem without our app?
- Where is the biggest friction point for trust, and how can we reinforce it?

Notes

1. This is a fictionalized story based on common user feedback regarding data consumption in Africa.
2. Intelpoint. "With a Median Age of 19 Years, Africa Has the Youngest Population, while Europe Has the Oldest Population (43 years)." Accessed February 24, 2026. https://intelpoint.co/insights/with-a-median-age-of-19-years-africa-has-the-youngest-population-while-europe-has-the-oldest-population-43-years/
3. Mobile Operating System Market Share Africa | Statcounter Global Stats. Accessed October 15, 2025. https://gs.statcounter.com/os-market-share/morbile/africa

4. Accessibility needs can differ by user or region. For example, Kome Sideso says accessibility needs in Kenya are more about content and language, while in Nigeria it's more about internet connectivity.

Bibliography

1. Mobile Operating System Market Share Africa | Statcounter Global Stats. Accessed October 15, 2025. https://gs.statcounter.com/os-market-share/morbile/africa
2. Intelpoint. "With a Median Age of 19 Years, Africa Has the Youngest Population, while Europe Has the Oldest Population (43 years)." Accessed February 5, 2026. https://intelpoint.co/insights/with-a-median-age-of-19-years-africa-has-the-youngest-population-while-europe-has-the-oldest-population-43-years/

Part 2

The APR™ in Context

Chapter 3

Building Mobile-First for African Contexts

In 2020, as Kenya went into COVID lockdown, a young psychology graduate named Beryl Njoki noticed a crisis that wasn't on the evening news. Schools had closed, and teenagers were home, unsupervised, restless, and scared. In the low-income neighborhoods around her, they were turning to cheap drugs and alcohol to cope with boredom, anxiety, and family stress.

"There was a pandemic within the pandemic," she would later say. Beryl didn't have a big grant or a Silicon Valley accelerator behind her. What she had was strong conviction and proximity: she was close enough to these young people to hear what they were going through, and they trusted her enough to say it out loud. So she started New Beginnings KWA Ground, a small organization offering an eight-session program called Kujali—"to care" in Kiswahili—for students, as well as sessions for parents and teachers.

Kujali was simple and very analog: group sessions in schools about emotions, coping, communication, family relationships, mental health, and substance use. Over time, New Beginnings

DOI: 10.1201/9781003648192-5

KWA Ground reached more than 3,500 students in four Kenyan schools and several communities.

But Beryl kept hearing the same thing from her students after each school term ended: “I wish there was a way we could still reach you when we go home.” Technically, they could. They had phone numbers for Beryl and her volunteers. But in reality, the model didn’t work. Calls came when she was in meetings or when volunteers were in class or out of network. Adolescents didn’t always feel brave enough to phone an adult about self-harm urges at midnight. And in many Kenyan homes, mental health is still something you are shamed for, not something you get help for.

The support that Kujali offered was strongest inside many school grounds and weakest precisely when young people needed it most: at home, on weekends, in the evenings, and during long holidays.

This gap pushed Beryl toward a digital product. When she and her team asked students what might help, the answers were concrete and very 2020s Africa: WhatsApp, voice notes, videos. In other words, the things they were already using every day on low-cost Android phones.

Out of that feedback came the idea for Kujali Chat Care: a WhatsApp-based chatbot designed as “a friend in the palm of your hand.” It’s not a doctor or a therapist. It checks in on your mood with emojis. It nudges you to think about what you’re feeling. It can pull in short lessons from the Kujali curriculum on coping skills, communication, peer pressure, bullying, or substance use, without sounding like a textbook. It “speaks” English, Kiswahili, and Sheng. It “knows” there are things a 14-year-old in Nairobi will never say to an adult in the same room but might type into a private chat at 11:47 pm.

Crucially, it also “knows” when to stop being “just a friend.” Behind the scenes, Beryl has had to design an escalation path: if a child talks about suicide, self-harm, or serious abuse, the system must route that conversation to a human, and sometimes from there into formal services.

Beryl didn’t “choose a WhatsApp chatbot” because it was trendy; she specifically chose it because it was the only realistic way to be available to her users in the channels they already used, with the devices and data they already had.

What "Mobile-First" Really Means in Africa

Beryl's story captures what mobile-first looks like in practice in sub-Saharan Africa. It's about designing for severe constraints related to connectivity, device capabilities, and user literacy, rather than simply optimizing a desktop site for a smaller screen.

Kenya, like Nigeria, South Africa, and Egypt, is a mobile-first country. For millions of people, a mobile phone (whether a smartphone or a basic feature phone) is their primary, and often sole, gateway to communication, financial services, education, and economic participation.

In many ways, this is simply the latest chapter in a long history of leapfrogging. Across the continent, countries skipped landlines altogether but rapidly adopted mobile phones and moved past traditional banking to embrace mobile payments at scale.

In 2024, Nigeria reached 85% mobile phone penetration, with Kenya and Egypt at 90% and 95%, respectively. (See Chapter 4, "Building for Digital Connectivity," for a deeper exploration of infrastructure and access.)

But those numbers hide a more complex reality. When we talk about "mobile phones," what exactly are people using?

FIGURE 3.1 Factors shaping what "mobile-first" means for Africa.

Device Access

Feature Phones

Despite poor connectivity and limited internet access, many Africans remain connected to digital services, thanks to feature phones.

While only 52 active mobile broadband subscriptions exist per 100 African residents, there are 98 active mobile-cellular telephone subscriptions. Furthermore, 66% of African residents aged 10 or older have a mobile phone compared to 38% with internet access.[1] In other words, people have greater access to calls and SMS—usually through feature phones, but sometimes also smartphones—than they have to wireless internet connections that enable browsing.

More than half of the mobile phones shipped to Africa in 2022 were feature phones.[2] These phones remain widely popular, especially in low-income communities, because they are cheaper, costing as low as 6$for one with internet features (2G), whereas the cheapest entry-level smartphones cost at least triple that price.[3] As GSMA notes, "In sub-Saharan Africa, …, an entry-level device costs 99% of average monthly income for the poorest 20%."[4]

Beyond affordability, feature phones are more durable and withstand falls and accidents better, have longer battery life (essential in contexts with limited electricity), and most of all, do not require expensive data and subscription packages.

The use of feature phones does not necessarily mean digital exclusion. Instead, it has prompted innovative, low-tech digital services that enable users to participate in the digital economy and access digital services. The most common example of this is USSD, which enables people disconnected from the internet to access financial services digitally. Functioning USSD services can also benefit smartphone users in areas with limited connectivity. While feature phones anchor connectivity, the market is gradually shifting toward more capable devices. The rise of smartphones introduces new opportunities, but also a complex set of hardware and economic constraints that product builders must navigate.

Smartphone Ownership

Smartphone penetration is increasing, though growth is not uniform across the continent. This shift is driven by various factors, including Buy Now Pay Later (BNPL) offers, Pay As You Go (PAYG) plans, an influx of affordable Android devices from Chinese manufacturers, and a strong second-hand device market.

As seen in Figure 3.2, Android devices dominate on the African continent.[5] iPhones, while popular in high-income urban areas, remain a small slice of the market due to cost. Much of Africa's smartphone growth has been driven by manufacturers like Transsion Holdings, the parent company of Tecno, Infinix, and Itel. The company consistently captures over 50% of the market by manufacturing functional smartphones priced under 100$ and tailored to the African market.[6] Transsion was among the first to introduce dual-SIM Android phones and promote longer battery life, features that address the digital behaviors and needs of many Africans.

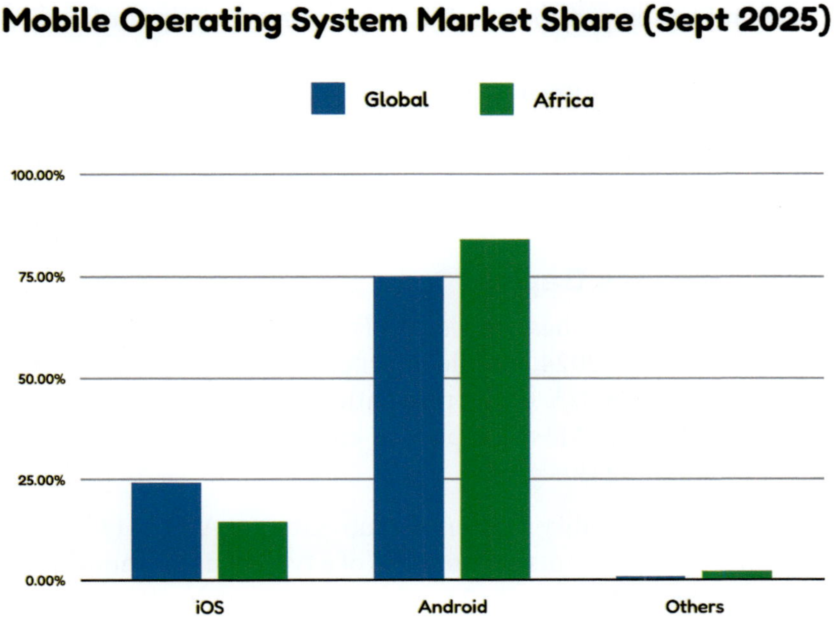

FIGURE 3.2 Bar chart showing the mobile operating system market share.

Economic factors also drive the market for used and refurbished smartphones, which cost a fraction of new smartphones.[7] While these second-hand devices expand access to the digital economy, they present challenges for the user experience. They have hardware limitations, such as poor battery life, limited storage space (which creates barriers to downloading apps and receiving or creating media), and lower screen resolutions resulting in poor display quality. These devices also typically have limited RAM (1–2 GB), outdated Android versions, and less powerful processors, affecting app size, performance, and compatibility. These all have implications for app size, performance, and compatibility. For example, users unable to update or download new apps due to storage limits may resort to cracked versions or sideloaded APKs, which pose security risks.

Second-hand devices received as hand-me-downs may have cracked or damaged screens or include apps left by the previous user, potentially hindering the new user's ability to properly onboard apps or fully experience their functionalities. Finally, tablet and laptop penetration is much lower, typically confined to urban professionals, students, or wealthier households. Yet this impressive reach does not translate into universal or consistent connectivity. A phone in someone's hand does not automatically mean reliable internet access or the skills and confidence to use it.

Digital Access Gaps

Sub-Saharan Africa has the world's largest mobile internet usage gap. By the end of 2024, mobile internet penetration reached only 27%, even though 87% of the population technically lived within network coverage. Most people are covered but not connected. Three forces drive this gap:

1. Data Affordability Barriers: In sub-Saharan Africa, 1 GB of data often accounts for up to 7% of a typical user's monthly income, well above the 2% affordability benchmark popularized by the Alliance for Affordable Internet (A4AI)[8]—this is especially burdensome for low-income households. By contrast, in many Southeast Asian markets (often used

as a reference point when designing for Africa), the same 1 GB may cost well under 1% of monthly income. In North America, despite higher nominal prices, the relative cost is a tiny fraction of income

2. Digital Skills and Literacy: Affordability alone does not explain the gap. Large segments of the population lack the digital skills or literacy required to navigate text-heavy applications, multi-step onboarding flows, or English-only interfaces. For many first-time smartphone users, even basic tasks, downloading an app and interpreting error messages, can be paralyzing without support. As a result, people often rely on voice notes, visual cues, or imitation rather than reading, which fundamentally shapes how products must be designed for these markets
3. Trust and Safety Concerns: Trust also operates as a powerful barrier. High exposure to fraud, impersonation, and digital scams has made many users wary of engaging online, particularly when financial information or identity verification is involved. Even when data and devices are affordable, fear of "doing the wrong thing" or being scammed can keep users offline or limit them to a narrow set of familiar apps

Device Sharing

Another key point regarding device use and access is the prevalence of phone sharing. Although ownership rates may be low, phone usage is considerably high as parents may share phones with children, husbands with wives, and friends with one another. As with the popularity of second-hand and low-end devices, this phenomenon is primarily, but not only, a result of economic constraints.

Because mobile phones have become a tool for economic advancement, phone sharing has become an important way to remain connected to essential services. In this case, it serves as an enabler of digital inclusion, social communication, and access. However, device sharing introduces privacy and data security issues that must be considered when making design and product decisions.

OPE ADEYEMI: A CASE STUDY ON REDESIGNING IDENTITY VERIFICATION FOR SHARED DEVICE CONTEXTS

One of the most significant cultural behaviors that forced a core design change in our onboarding and KYC products was the use of shared devices and collective account setup in informal markets. In many peri-urban and rural communities across Nigeria, a single smartphone is often shared among multiple users, sometimes an entire household, and in business clusters, by several merchants who rely on one device to access financial services.

We discovered this pattern through field research and escalation trends. When we investigated further through merchant interviews and agent ride-alongs, we learned that many customers depended on their Business Relationship Managers (BRMs) to open or manage their accounts using the BRM's own phone. While this behavior was socially convenient, it severely undermined identity integrity, KYC accuracy, and transaction accountability.

We addressed this by redesigning our onboarding architecture to de-couple identity verification from the device while still respecting local realities. First, we introduced device-binding with biometric consent, requiring each customer to verify ownership of a device before critical actions such as login. Second, we implemented multi-device session management, allowing secure logins from different devices without compromising risk controls. Finally, we rolled out face re-verification on suspicious sessions, using our fraud intelligence layer to detect anomalies in login patterns based on signals such as distance from trusted location and transaction activity.

Beyond economic factors, cultural practices and norms also drive phone sharing, although the practice is not always voluntary. Another factor driving shared devices is digital literacy. People with lower levels of digital literacy may rely on a more digitally literate person to manage their digital activities, such as sending and receiving text or WhatsApp messages or conducting internet or mobile banking transactions. These dynamics often involve power imbalances[9] between parents and children, or between husbands and wives, whereby one party acts as "gatekeepers" of the others' devices.

Beyond phone sharing, devices may be shared in several other contexts. These include accessing a shared computer in a public space, such as a cybercafe or library, or for work purposes, such as in hospitals, labs, hotels, and so on.

JOSHUA ADENUGA: A CASE STUDY ON ALIGNING DIGITAL PRODUCT FLOWS WITH REAL USER BEHAVIOR

Joshua Adenuga, Head of Product at FPG Technologies & Solutions, encountered a pivotal design challenge while developing CivisHealth, an AI-powered Health Management Information System (HMIS) for hospitals and clinics across Nigeria. The single biggest cultural behavior that prompted his team to change their product design was the widespread sharing of computers among healthcare staff, combined with the practice of delaying patient documentation until much later in the day.

During field visits and pilot deployments, they observed that most hospitals had only a few desktop computers for entire wards. Doctors used them collectively for consultations, billing, and reporting. During ward rounds, they left the systems behind, took handwritten notes, and later returned to type summaries into the system. Due to heavy workloads and long outpatient queues, many entries were shortened or skipped entirely.

At first, the team thought this was simply an infrastructure limitation. However, after shadowing clinicians and interviewing hospital administrators, Joshua realized it was a cultural pattern built around shared access and time pressure—and that it had serious implications. Delayed or incomplete entries led to inaccurate patient records, revenue leakages, and frequent audit disputes between hospitals and Health Maintenance Organizations (HMOs). In other systems, staff could even edit consultation data days later, often after claims had already been submitted, creating mismatched records that HMOs refused to honor.

To resolve this, Joshua made a major product decision: extend CivisHealth to mobile devices so that each healthcare worker could log in securely from their own phone using a personal ID. This allowed doctors and nurses to capture consultation notes, vitals, and prescriptions in real time during outpatient engagements and ward rounds. It eliminated their dependence on shared desktops and made documentation part of their normal workflow.

At the same time, the team introduced strict edit controls to preserve trust and accuracy. Consultation records can only be edited within 24 hours and only by the original creator. Lab and radiology requests are locked immediately after submission. These measures ensured audit integrity and ended the long-standing conflicts between hospitals and HMOs over altered data or date mismatches.

This change fundamentally shifted staff behavior: documentation became timelier, audit trails remained intact, and HMO claim rejections dropped sharply. Hospital administrators gained confidence in their digital records, and clinicians reported that the system finally "fit how they actually worked."

By addressing this single cultural behavior—shared device use and delayed entry—Joshua's team strengthened accountability and trust across the entire healthcare ecosystem. It taught them that in Africa, good product design must often adapt to human patterns before humans can adapt to technology.

Design Implications of Africa's Mobile-First Reality

To build something inclusive and accessible for sub-Saharan Africa, you need to think about two types of constraints simultaneously:

A device-first approach starts with the realities of the hardware itself. It assumes a small screen, intermittent connectivity, battery anxiety, and low-end devices. In practice, that often means designing for a phone with around 16 GB of storage—much of it already used—running an older version of Android and losing connection without warning. Products built for this environment need to stay lightweight and continue working on hardware that would be considered outdated in many Western markets.

Device constraints show up most clearly in decisions about apps and codebases. Many services choose to be accessible only through app stores like the Google Play Store or Apple App Store, sometimes with a limited web version as an afterthought. This decision makes sense from a business perspective—downloads are an easy metric to track. But from a user's perspective, installing an app is a commitment because space is limited and updates consume data.

Insisting on app-only access can quietly exclude people whose phones cannot accommodate large downloads or frequent updates, or who are cautious about installing unfamiliar software. In these contexts, Progressive Web Apps (PWAs) and Lite versions are not secondary experiences. They are often the difference between being able to try a service at all and being locked out entirely. They reduce storage pressure, lower data costs, and allow people to engage without committing scarce resources upfront. PWAs also have their limitations, so it is important to weigh the benefits and costs. For example, they may consume more battery power because they rely on browsers as an intermediary to work, and they are not ideal for complex and resource-heavy tasks.

A context-first approach recognizes that phones are often shared; access may be shaped by parents, spouses, or employers; and local norms influence what feels safe to open, search, or save. The person using your product may not own the device. They may be navigating stigma around certain topics or relying on children, neighbors, or more digitally confident relatives to help with anything that feels complex or risky.

CONTEXT-FIRST REALITY: DESIGNING FOR SHARED DEVICES

Hillary co-led discovery research for a local-language feature for a global messaging app in sub-Saharan Africa. Going in, she assumed she understood the region well. As a Nigerian who had lived in Ghana for several years, made multiple visits to Kenya, and conducted extensive desk research and literature reviews, that confidence felt justified. But she quickly realized how incomplete her understanding was.

Across research conducted in Ethiopia and Nigeria, many participants used budget Android devices, which typically had limited storage, inconsistent processing power, and older operating systems.

Beyond device limitations, there was also the constraint of shared devices. Women and young people, in particular, often relied on devices owned by parents, siblings, or other relatives. High mobile penetration figures are often interpreted as a proxy for individual ownership. In practice, they don't account for shared access, which creates a very different set of design conditions.

Uncovering Hidden Needs

One of the most valuable benefits of rigorous discovery work and user research is identifying the gap between perceived and actual competence—what the Dunning-Kruger effect describes. Participants believed they were experienced WhatsApp users, but their actual proficiency was quite limited. They were locked into basic usage patterns, unable to discover features that required reading text, unable to troubleshoot problems independently, and constantly fearful of making mistakes when they encountered unfamiliar screens.

This pattern reflects a broader reality in mobile-first contexts. People will continue using a product if its core value is clear and accessible, but they may avoid large parts of the experience while doing so. Over time, they learn which elements feel safe, which are likely to cause confusion, and which should be ignored altogether.

This learned avoidance became especially visible during testing of a language-selection feature. When language options were presented as secondary elements on a welcome screen, most participants overlooked them entirely. Their attention immediately went to the primary call to action. Past experience had taught them that secondary links on welcome screens—terms, privacy policies, settings—were usually written in English and rarely relevant.

Even when participants noticed the language option, many chose not to engage. Some assumed it was a static label rather than a setting. Others expected it to open a keyboard menu or believed it would contain only international languages unrelated to their own. After years of being underserved, they did not expect to see Hausa, Amharic, or Afaan Oromo represented at all.

When language selection was instead surfaced as a full-screen step before any terms or agreements, with clear icons

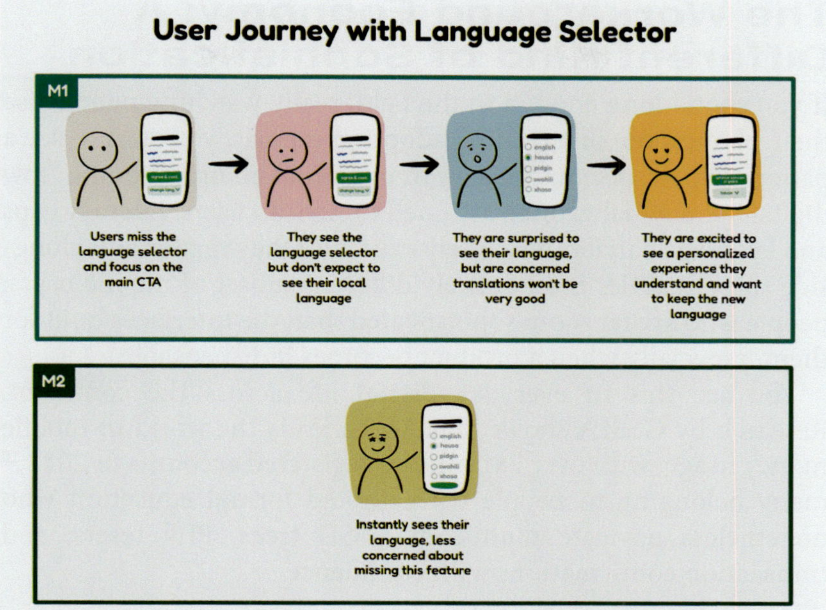

FIGURE 3.3 User journey with language selector.

and local languages prominently listed, behavior shifted. Users engaged confidently and completed setup with less assistance. More importantly, the product sent a clear signal: their needs were anticipated, not treated as an edge case.

What This Reveals for Mobile-First Design

Mobile-first design in African contexts cannot stop at screen size, performance, or bandwidth optimization. It must account for shared devices, uneven literacy, learned avoidance, and the social cost of making a mistake. Elements that appear secondary or optional to experienced users may be actively avoided by others for good reason.

Designing for inclusion often requires rethinking where critical decisions appear, how language is surfaced, and which assumptions are safe to make. Context-first design is the difference between access in theory and access in practice. Yet constraints alone do not explain how people behave.

The Workaround Economy: A Different Kind of Sophistication

If you spend long enough in the field, really watching people use their phones, not just reading adoption reports, you start to see a pattern with a specific category of users that complicates the "low digital literacy" label often assigned to African users. Literacy gaps and language barriers absolutely exist, and they matter. But alongside these realities is an entirely different subset of mobile users: people who are far more sophisticated than the interfaces built for them, especially when a product becomes indispensable.

You see this in everyday digital life across the continent. Research by GSMA shows that Africa leads the world in mobile money usage, with over 781 million registered accounts in 2022,[10] many belonging to people with limited formal education who nonetheless navigate multi-step USSD trees, PIN resets, and transaction confirmations with confidence.

WORKAROUNDS FOR DIGITAL LITERACY AND LANGUAGE

Language barriers compound the challenges of using digital apps. This same ingenuity surfaced clearly in the messaging app study. Many participants struggled to download and set up mainstream apps without assistance.

Registration flows, permission requests, and system messages were often presented only in English, making it difficult to understand what information was being requested or why an action had failed. Some participants described attempting to register independently but being unable to proceed without help from a friend or neighbor.

To navigate these barriers, people developed workarounds. These included asking others to translate screens one by one, switching between multiple keyboards, or cross-checking unfamiliar words in dictionaries. While these strategies demonstrated persistence and adaptability, they also placed a heavy cognitive burden on users for tasks that designers typically assume are simple.

Not everyone could sustain that effort. For each person who managed to push through with help, others were excluded

FIGURE 3.4 Digital literacy is sometimes built through informal guidance from others.

entirely because they lacked confidence, time, or someone nearby to assist them. This is how products that appear usable in aggregate quietly fail large segments of their intended audience.

Once participants gained access with assistance, many settled into a narrow but functional mode of use. They relied heavily on voice notes and emojis, memorized the layout of the interface, recognized icons by position rather than label, and learned a small set of gestures that allowed them to communicate without reading. This created a sense of competence, but it was fragile. Interface changes, unexpected prompts, or new flows could quickly undermine that confidence.

These workarounds reveal that traditional literacy is not always correlated with digital literacy. Age and formal education are not necessarily proxies for digital capabilities. Field research by Yann Le Beux of YUX Design exemplifies this. In Senegal, where formal French literacy is relatively low, they found urban users to be exceptionally proficient with mobile devices, navigating complex digital tasks with ease. Whether it's a market trader in Dakar or a motorcycle taxi driver in Nairobi, these users have taught themselves to navigate complexity through motivation and pattern recognition rather than traditional textual literacy. As Yann notes,

"Seeing this disparity helps you focus on segmentation based on actual behavior rather than just educational status."

This behavioral insight has caught many product teams off guard. According to Paul Otu, Head of Design Strategy at Interswitch Group, his team experienced this while building a consumer payment solution for lower-income communities in Nigeria. Their starting point felt reasonable: these users have limited digital literacy, so strip everything back and build the most basic, bare-bones interface possible. But when they started looking at what apps their target users had on their phones, the reality didn't match the assumption. "These 'low digital literacy' users were confidently navigating some of the most complex apps in the market," Otu explained. "Betting platforms, microloan apps, multi-level marketing platforms—apps that even we, as designers, sometimes find confusing."

Most sports bettors have only a secondary school education or less, yet they regularly handle nested menus, odds tables, real-time updates, and dense information structures on these platforms. By every general design principle, these interfaces should have been unusable for this demographic, but they weren't. These users had simply taught themselves how to make it work. They created their own mental models—ones that didn't depend on literacy so much as motivation and pattern recognition.

Otu's team ultimately reframed their whole approach after this insight: "We stopped asking 'How simple can we make this?' and started asking 'Is our value proposition strong enough to justify what we're asking users to invest?'"

Busolami Kehinde, who has spent over half a decade building consumer apps, noticed a similar pattern when she worked at a fintech. Users repurposed the crypto funding feature as a kind of arbitrage play, repeatedly funding and withdrawing to exploit tiny windows of value and never touching the long-term investment product the team assumed they were there for. This behavior might look irrational from a product lens, but it makes perfect sense when viewed through the lens of economic self-optimization. It mirrors broader findings in a joint research by Caribou Digital and Mastercard Foundation, which documented how African users routinely repurpose digital financial tools in ways product teams never anticipated.[11] We see this pattern too. People using savings apps as remittance rails, cashback features as income sources, and crypto tools as short-term liquidity mechanisms.

Across all these examples, a core theme emerges: "When motivation is strong enough, users in these markets become high-context learners who can absorb complexity through observation, experimentation, imitation, and social learning."

For features with a clear, immediate payoff, such as checking your balance, sending money, and making money, users will tolerate meaningful complexity if it leads to better outcomes. For low-stake features like profile updates or personalization, brutal simplification still matters. And for everything in between, the answer is progressive disclosure: start simple, then reveal depth as users engage and signal that they're ready for more.

The Rise of Super Apps

This same logic—users conserving cognitive effort for apps that have proven their value—explains another major pattern: the rise of super apps across Africa.

Instead of asking users to download, learn, and trust multiple apps, each with its own interface logic and potential for error, like their Southeast Asian predecessors (WeChat, Alipay, Grab), African super apps such as M-Pesa, OPay, and Gozem[12] thrive because they combine many use cases into a single, lightweight container.

Sub-Saharan African smartphone users tend to install and actively use relatively few apps compared with global averages, constrained by limited device storage and data costs on low-end devices. Moreover, each new app represents additional mental overhead—another interface to decode, another trust decision to make.

Super apps reduce this cognitive load. Once users have invested the effort to learn M-Pesa's logic or OPay's interface, adding bill payments or ride-hailing ends up being an incremental adjustment rather than starting from zero. FSD Africa research suggests that new financial services are more likely to be adopted when delivered through a platform or service channel they already trust.[13] Essentially, it's not that people want everything in one app—it's that once trust is earned, it becomes the doorway through which they're willing to explore more.

What we realize from these stories and being grounded in real discovery work is that mobile-first design in Africa grows out of a thousand tiny negotiations happening in people's pockets: how much data they can spare today, which apps feel safe to open in public, who else might read their messages at home, whether the network will hold long enough to complete a task.

You see these negotiations shaping how people use technology, not the idealized version we sketch on whiteboards or brainstorm in Zoom meetings. You start to see how quickly things fall into place when a product respects those realities. This is what allowed tools like Beryl's chatbot to work: it fit into an existing rhythm. Everything that follows builds on this simple truth: understanding the everyday context is the beginning of good design.

Key Takeaways

1. The primary design challenge is around non-hardware barriers like data cost, digital literacy, and lack of trust, so to truly design for mobile first, you have to consider context first.
2. Design must account for the reality of widespread device sharing, which requires building in strong privacy, security, and personal login flows.
3. Product success can be enhanced by consolidating trust, that is, reducing the cost and risk of adoption by consolidating services under a single, already-trusted platform.

What Next?

The Constraint Audit (Exercise/Checklist)

What Are You Actually Designing For?

For your current or next product targeting African users, complete this audit honestly:

Device Reality Check:

- What's the median storage capacity of devices your target users have? (If you don't know, how will you find out?)

- What's your current app size? How does it compare to the available storage on your users' devices?
- Does your app work smoothly on a three-year-old Android device with 2GB RAM?

Data and Connectivity:[14]

- How much data does your onboarding flow consume? Your core user journey?
- What happens when connectivity drops mid-task? Have you tested this?
- Do you have robust offline functionality, or is it on your "nice-to-have" list?
- What percentage of your target users are on 3G versus 4G/5G?

Context and Access:

- What percentage of your users likely share devices with family members?
- What are the privacy/safety implications of shared devices for your product?
- In what languages is your interface available? Is that based on data or assumptions?
- Can users complete core tasks through observation and pattern recognition, or do they need to read everything?

Value Proposition:

- Is your value proposition strong enough that users will tolerate friction to get to it?
- Are you asking for permissions/information before you've proven value?
- What's the cognitive cost of learning your interface? Is it justified by the payoff?

Research and Validation:

- When did you last observe actual users completing tasks in your app?

- Does your sample reflect the diversity of your actual user base (devices, literacy levels, contexts)?
- What workarounds have you observed users creating? What do they tell you?

Most importantly: Do real discovery work to understand the actual experiences of your target or current users. Watch them use your product in their real contexts, not in a lab, and certainly not just through analytics dashboards, but on their devices and in their daily lives.

Notes

1. "Facts and Figures 2024: Mobile Phone Ownership," International Telecommunication Union (ITU). November 10, 2024. https://www.itu.int/itu-d/reports/statistics/2024/11/10/ff24-mobile-phone-ownership
2. Ecofin Agency, "Africa: Five Countries Have Near 100% Mobile Phone Ownership (Survey)." Last modified January 16, 2023. https://www.ecofinagency.com/telecom/1601-44176-africa-five-countries-have-near-100-mobile-phone-ownership-survey
3. Chris Donkin, "Feature: What Is Hampering the African Smartphone Market?," *Mobile World Live*. August 14, 2024. https://www.mobileworldlive.com/devices/feature-what-is-hampering-the-african-smartphone-market
4. GSMA, *The State of Mobile Internet Connectivity 2024* (GSMA 2024), 5. https://www.gsma.com/r/wp-content/uploads/2024/10/The-State-of-Mobile-Internet-Connectivity-Report-2024.pdf
5. Mobile Operating System Market Share Africa | Statcounter Global Stats. Accessed October 15, 2025. https://gs.statcounter.com/os-market-share/morbile/africa
6. Canalys, "Africa's Smartphone Market Grows 7% in Q2 2025, Marking Nine Straight Quarters of Growth," *Omdia by Informa*. Archived August 27, 2025 at https://web.archive.org/web/20250827130850/https://www.canalys.com/newsroom/africa-smartphone-market-q2-2025
7. Manish Pravinkumar, "Rising demand for refurbished smartphones in the Middle East and Africa (part 1)," *Canalys* (now *Omdia by Informa*). Archived June 5, 2024 at https://web.archive.org/web/20240918045216/https://www.canalys.com/insights/rising-demand-for-refurbished-smartphones-in-mea-part-1
8. Alliance for Affordable Internet (A4AI), "Affordable Internet Is '1 for 2'". Accessed November 24, 2025. https://a4ai.org/affordable-internet-is-1-for-2/

9. See Chapter 5 on cultural classifications for more on this.
10. Chris Green, "GSMA's Mobile Money Report: 17% Growth and 781 Million Accounts in Africa," *TechAfrica News*. April 19, 2023. https://techafricanews.com/2023/04/19/gsmas-report-on-mobile-money-sheds-light-on-the-rising-popularity-of-the-service-globally/
11. Partnership for Finance in a Digital Africa, *How Do Clients Use Digital Finance in Daily Life and Daily Practice?* (Caribou Digital, 2017), 4. https://caribou.global/publications/how-do-clients-use-digital-finance-in-daily-life-and-daily-practice/
12. African super apps such as M-Pesa (Kenya's mobile money platform), OPay (Nigerian fintech), and Gozem (West African ride-hailing and delivery) combine start multiple services—payments, transfers, savings, loans, transport, and commerce—into single platforms.
13. Elizabeth Kiamba, "FSD Network Collaborations Aimed at Harnessing the Power of the Digital Platform Economy," *FSD Africa*, November 18, 2019, https://fsdafrica.org/fsd-network-collaborations-aimed-at-harnessing-the-power-of-the-digital-platform-economy/
14. More in Chapter 4.

Bibliography

Alliance for Affordable Internet (A4AI). "Affordable Internet Is '1 for 2.'" Accessed November 25, 2025. https://a4ai.org/affordable-internet-is-1-for-2/

Canalys. "Africa's Smartphone Market Grows 7% in Q2 2025, Marking Nine Straight Quarters of Growth." *Omdia by Informa*. Archived August 27, 2025, at https://web.archive.org/web/20250827130850/https://www.canalys.com/newsroom/africa-smartphone-market-q2-2025

DataCup. "How Has Mobile Internet Penetration Developed in Africa?" Accessed November 25, 2025. https://datacup.io/en/blog/datacup-1/how-has-mobile-internet-penetration-developed-in-africa-40

Donkin, Chris. "Feature: What Is Hampering the African Smartphone Market?" *Mobile World Live*, August 14, 2024. https://www.mobileworldlive.com/devices/feature-what-is-hampering-the-african-smartphone-market

Ecofin Agency. "Africa: Five Countries Have Near 100% Mobile Phone Ownership (Survey)." Last modified January 16, 2023. https://www.ecofinagency.com/telecom/1601-44176-africa-five-countries-have-near-100-mobile-phone-ownership-survey

Green, Chris. "GSMA's Mobile Money Report: 17% Growth and 781 Million Accounts in Africa." *TechAfrica News*, April 19, 2023. https://techafricanews.com/2023/04/19/gsmas-report-on-mobile-money-sheds-light-on-the-rising-popularity-of-the-service-globally/

GSMA. *The State of Mobile Internet Connectivity 2024*. GSMA, 2024. https://www.gsma.com/r/wp-content/uploads/2024/10/The-State-of-Mobile-Internet-Connectivity-Report-2024.pdf

International Telecommunication Union (ITU). "Facts and Figures 2024: Mobile Phone Ownership." International Telecommunication Union (ITU). November 10, 2024. https://www.itu.int/itu-d/reports/statistics/2024/11/10/ff24-mobile-phone-ownership

Kiamba, Elizabeth. "FSD Network Collaborations Aimed at Harnessing the Power of the Digital Platform Economy." *FSD Africa*, November 18, 2019. https://fsdafrica.org/fsd-network-collaborations-aimed-at-harnessing-the-power-of-the-digital-platform-economy/

Mobile Operating System Market Share Africa | Statcounter Global Stats. Accessed October 15, 2025. https://gs.statcounter.com/os-market-share/morbile/africa

Partnership for Finance in a Digital Africa. *How Do Clients Use Digital Finance in Daily Life and Daily Practice?* Caribou Digital, 2017. https://caribou.global/publications/how-do-clients-use-digital-finance-in-daily-life-and-daily-practice/

Pravinkumar, Manish. "Rising Demand for Refurbished Smartphones in the Middle East and Africa (part 1)." *Omdia by Informa*. Archived June 5, 2024, at https://web.archive.org/web/20240918045216/https://www.canalys.com/insights/rising-demand-for-refurbished-smartphones-in-mea-part-1

Chapter 4

Building for Limited Digital Connectivity

Once, while visiting a remote village in Nigeria, Lade was essentially disconnected from the internet for four days. Nothing worked. She could barely use WhatsApp or check her emails and couldn't watch YouTube or stream anything. She was not prepared for this. More recently, she was traveling by train from one Western European city to another. She was preparing a presentation she'd be giving on "Technological Accessibility", but the connection was so poor that she couldn't do much online. On the other hand, she was able to continue working on the presentation itself because Google Slides allows you to work offline and automatically synchronizes the content once you are reconnected to the internet.

These two experiences show the importance of designing for limited connectivity situations. When design choices and product decisions are being made, whether in Berlin, London, or San Francisco, little thought is given to whether people can actually access these digital products. Internet connectivity is treated as a given—moreso, perfect, always-on internet connection. But this is not always the case.

DOI: 10.1201/9781003648192-6

FIGURE 4.1 A woman working on the train in Europe versus a woman struggling with the internet in the village.

Landscape of Digital Connectivity in Africa

The International Telecommunication Union (ITU), a United Nations agency, reports that although the global offline population is steadily declining, the numbers remain low in Africa—only 38% of Africans use the internet, compared to over 87% in Europe and the Americas.[1]

When discussing the connectivity landscape, it is important to differentiate between coverage and usage. Coverage refers to whether or not a broadband connection is available, while usage refers to whether the available coverage is in use. Both coverage and usage access are skewed based on various demographic factors, including age and gender. Rural areas have less internet coverage and usage than urban areas, and women have less access than men.[2]

While the coverage gap has been reducing year over year from about 46% in 2015 to 13% in 2023,[3] connectivity is still not great.

Three of the Big Four (Nigeria, Kenya, and South Africa) rank among the bottom 50 in the global speed index.[4] GSMA estimates that, as of 2023, 2G and 3G networks make up about 70% of total connections in sub-Saharan Africa, while 5G makes up less than 1%, though it is estimated to grow to 17% by 2030.[5] 2G and 3G networks have wider and more affordable coverage for remote and rural areas, and they support basic services, such as SMS and USSD, that are essential for economic and social participation and are easily accessible on feature phones.

The usage gap shows the opposite trend: it has increased from around 42% in 2015 to 60% in 2023.[6] This means that even as mobile internet connectivity is growing in sub-Saharan Africa, the proportion of people actually accessing and using it has been declining over time. In other words, people are not using mobile internet even though they live in connected areas.

The variation in internet speed, the coverage gap, and the usage gap are attributed to several factors: underdeveloped infrastructure, high operational costs for telecommunications companies, high device costs, expensive data subscriptions, and limited digital literacy levels.

Underdeveloped network infrastructure hits hardest in remote and rural areas, resulting in poor coverage. Telecommunication companies also face operational challenges that limit their capacity to provide coverage or make it expensive to do so. Poor power supply, for example, forces companies to use diesel generators, which incur high fuel and maintenance costs. This drives up operating costs, which in turn widens the coverage gap.

The biggest driver of the usage gap is the high cost of data subscriptions relative to income. People simply can't afford to be (always) connected to the internet, even when there is coverage. On average, Africans spend about 4% of their monthly income on 1GB of data, which is twice the global target.[7] Of 13 regions, sub-Saharan Africa has the fourth-highest data costs, with half of the 10 most expensive countries located there.[8] One factor driving costs up is that people often buy smaller daily data amounts, which add up to be more expensive than monthly subscriptions in the long run. However, data costs can vary widely within the continent. Kenya and South Africa, for example, have more advanced infrastructure, resulting in 1GB data costs below the global average.[9]

Lade's research at an internet technology company showed that users would turn off data to manage it because of the high data costs. They often stay offline intentionally until it is absolutely necessary or they can afford airtime. Users going offline doesn't mean they have churned or are no longer interested in using the app.

Consequently, people often use multiple SIMs from different service providers as they shop around for cheaper services and more reliable connections, depending on their location. This behavior is evident in the disparity between mobile phone subscription rates and individual phone ownership.

According to Pew Research,[10] in addition to problems getting a reliable connection, mobile phone users in emerging economies take active steps to conserve their data, including avoiding certain activities they think might consume too much data. They also rely on low-data apps and tend to avoid high-data activities when there is no access to Wi-Fi. This drives preference for applications and services optimized for low data consumption, such as Facebook Lite.

Because of poor connectivity and high data costs, these users are often streaming or downloading low-quality videos. It also creates challenges in participating in digital activities like remote work and remote education, joining virtual conferences, and downloading educational materials.

It is also common for people to seek free public Wi-Fi, usually in or around banks, universities, transportation hubs, parks, malls, libraries, and other public buildings. For years, governments have worked with private organizations and nonprofits to create free Wi-Fi stations and zones to improve internet access and usage, often as part of digital inclusion strategies[11]. Initiatives across Africa, including in Nigeria, Rwanda, South Africa, Kenya, Malawi, and Zambia, aim to reach those who may be primarily offline due to cost constraints.

For some users, free Wi-Fi may be the only way to connect to the internet most, if not all, of the time, and these free Wi-Fi stations help people manage their data. Someone could wait until they are in a free Wi-Fi zone to engage in data-intensive tasks, such as updating apps, downloading large documents or videos to watch later, or streaming content. This behavior of delaying

FIGURE 4.2 Woman turning off her mobile data before going to bed.

data-heavy activities creates predictable usage patterns that product designers can anticipate and design for.

However, free Wi-Fi does not entirely solve the access problem. Daily allocations are often limited, sometimes as low as 50MB.[12] Some users may not be able to take advantage of these resources because of limited access to devices like smartphones and laptops. Additionally, many initiatives may not be sustainable due to poor management, limited funding, and changes in government.

Beyond seeking Wi-Fi, many users engage in workarounds to save on data costs. Yann Le Beux observes that sharing apps and content via Bluetooth is still a common practice.

Why Should You Still Build for Africa?

Despite the challenges outlined in this chapter, Africa remains a market with immense potential.

Internet penetration has been trending upward in Africa, with a 115% increase in internet users between 2016 and 2021.[13] This growth is expected to continue increasing, given the youth population. In 2024, an estimated 53% of young people aged 15–24 used the internet compared to 34% of those outside that age range.[14]

Furthermore, although these percentages appear lower than those of other regions, the raw numbers are massive. Nigeria, Kenya, South Africa, and Egypt collectively have approximately 281.5 million internet users as of the time of writing.[15] For context, the total estimated US population as of this time is just over 342 million. As internet penetration rates rise, the number of people participating in the digital economy will also increase.

This creates an opportunity for global businesses looking to expand or homegrown businesses looking to scale. Designing for the current realities highlighted thus far can create a powerful long-term advantage: as coverage and usage improve, users already accustomed to and loyal to a particular service are less likely to switch. This is ecosystem entrenchment in action. Companies like Google and Facebook used this strategy in emerging markets by building products around unique local challenges, to the point that their services have come to sum up the internet for many users.

By creating useful and usable products based on today's context, you are building a user base that is likely to grow with you, giving you an advantage over competitors who enter the market later. For example, high data costs, a major driver of the usage gap, necessitate designing data-efficient applications. When these services and products integrate seamlessly into the customer context, the mental and financial cost of switching could help you create a "sticky" ecosystem.

Finally, designing for so-called "edge cases" always has broader impact and implications. Many commonly used digital and non-digital tools and devices were developed to address accessibility needs. For example, electric toothbrushes were first created to make it easier and more efficient for people with limited motor skills to brush their teeth daily. Now they have become mainstream consumer products. Similarly, Alt Text was initially designed to make web content accessible to people who use screen readers, but it now aids anyone when images don't load properly due to poor connectivity and improves search engine optimization.

Designing for accessibility and usability in the context of digital access and literacy can also be applied to improve other products or operate in other markets, providing a competitive advantage.

Implications on Digital Experiences and Expectations

In contexts with limited access, "usability" is more about whether a service can provide value reliably, even with minimal data and poor latency. Now that we have established the landscape and the importance of designing usable systems for African users, let's explore the practical design, product, and business implications. The goal is to give you the background for understanding user expectations and behaviors in the context of digital products.

Data Consciousness and Transparency

Because of high data costs and poor connectivity, efficiency, data conservation, and reliability are key concerns for internet users in this region. Many users are keenly and constantly aware of their data balance, treating it as a finite resource to be conserved and managed.

To minimize costs, users employ data management techniques and behaviors, such as turning off data when not in use to prevent background consumption by apps and refusing to download apps or files if they don't know how much data it would cost them.

Complex animations, video explainers, and large, heavy images can cause users to drop off. Instead, keep it simple and display data usage information clearly. For example, show the size of media and file downloads, and give users control over media and attachment downloads rather than initiating automatic downloads (as seen in WhatsApp and Telegram).

FIGURE 4.3 Give users clear control over media attachments and downloads.

Offline First

An offline-first strategy[16] is another application of the "edge case is the best case" principle. Enabling apps to function offline and sync when a connection is reestablished allows users to benefit from using the app even without connectivity. When users can save information, access content, or perform actions locally without an internet connection, they are not automatically limited or excluded from using the product or service.

For example, on WhatsApp, you can type and send messages without an internet connection, and they are delivered once the connection is restored. Netflix and YouTube let users download videos to watch later. These features allow users to get value from services even when they lack immediate internet access.

When Lade worked with a global browser company to conduct research on improving offline experiences for Android users in sub-Saharan Africa, the goal was to identify how the browser could provide better value to users navigating limited connectivity. The research revealed that "saving" content was a nuanced behavior. Users were "bookmarking" via workarounds—taking screenshots or sending links to themselves via SMS and WhatsApp. Users strategically planned their connectivity. They would wait to open these links until they reached a location with free Wi-Fi, such as a workplace or school. Eventually, the team developed a "Choose when to download" feature which aligned with the user's existing behavior of waiting for "free" internet by allowing them to select whether to download a file "Now," "On Wi-Fi," or at a specific "Date and time."

Services that require constant connection to function may have lower engagement and adoption rates than those that don't. The practice of using free Wi-Fi stations for data-heavy activities also reinforces the point. Not having access to the internet now does not mean being entirely cut off, and digital products should support existing user behavior.

Clear Feedback

Designing for offline first means designing for all scenarios—including partial loading, long or indefinite wait times, and complete failure scenarios—by providing clear status feedback. Many users understand and may even expect poor connectivity, so they are more patient with loading screens and delays, but there must be clear feedback that lets them know the connection status. Even when data transfer is delayed, clear messages and loading placeholders can ease the user's frustrations, salvaging what might otherwise be a bad experience. This is where good UX Writing also comes in: proper error handling includes user-friendly messages on what's happening and what users can do instead.

Offline Access

Another offline-first approach could be providing access to essential features of your service or product through alternate means.

Just as financial services providers enabled participation in the digital economy by allowing users access to services using USSD, businesses in other sectors can do the same.

Offline-first design is not only about supporting the app when the user is not connected to the internet but also about how they can access basic functionalities. If you can't log in to your internet banking or without a smartphone, for example, you can still check your balance using USSD. In Senegal, Wave is offering a non-USSD, offline-digital bridge by issuing small physical cards—a piece of paper, often the size of a business card—with a QR code printed on it. This card acts as a mobile money wallet, and the user can give the QR code to vendors or agents and participate in the digital economy.

When developing these systems, however, security concerns need to be addressed, especially around authentication. For example, if someone's phone is stolen, an unauthorized person could transfer money out of their account using USSD.

Mobile money platforms like M-Pesa in Kenya illustrate how to adapt to the local context of limited connectivity through offline-first design and "the edge case is the best case" approach. M-Pesa achieved mass adoption by designing for the edge case: a simple feature phone with a basic connection. This focus on the "worst case" scenario of slow, expensive, or no internet connection facilitated financial inclusion at scale.

Efficient Data Transfer and Content Optimization

Offline-first design also includes reducing API call frequency and payload size, implementing caching strategies, using efficient data formats, and prioritizing critical information to load first. These are all direct responses to the need for optimized code bases and smaller service payloads.

EXPERT INSIGHT: TOYIB MUSTAPHA ON DESIGNING FOR TECHNICAL RESILIENCE AND DATA EFFICIENCY

As a software engineer, my core focus is technical efficiency. This means embracing techniques like caching and state management. For instance, a user's basic details (like username or profile picture) are stored locally on login and managed as part of the application's state. This prevents unnecessary, simultaneous API calls and reduces the constant burden on a low-speed internet connection to fetch the same data repeatedly across different screens.

For handling offline periods, I implement robust offline support—much like how messaging apps queue unsent messages. Any critical user action or data submission is queued locally and automatically synced to the API or database once a stable connection is re-established. This allows the product to remain usable and manage data transfer reliably even with inconsistent service.

It is important to handle user experience locally as much as possible to minimize database roundtrips. When dealing with large assets, content optimization is essential:

- Ensuring we serve the right resolution of images to load quickly
- Employing techniques like lazy loading to prioritize the most important content
- For uploads, especially large files, we send data in small, segmented batches, rather than as a whole, often by encoding and uploading in "droplets." This, combined with setting a maximum size limit for files, significantly improves upload success rates on unstable networks

Efficient content delivery mechanisms can improve the experience when there is limited connectivity. For images, this means using the right optimization techniques to aid image compression and responsive image delivery, using appropriate image formats, and using placeholder images.

The same principle applies to video and audio. Media autoplay should be avoided because of data-conscious users. Furthermore, image and video compression that maintains quality is essential in low-data contexts, as is the use of lighter file formats and optimized code to reduce overall service payload.

For video optimization, adaptive bitrate streaming adjusts video quality based on current conditions, ensuring a smoother viewing experience. This is far better than forcing high-quality video streams that consume much data or fail to load. Transcoding video to lower resolutions also caters to the capabilities of entry-level smartphones with less powerful processors by creating smaller versions of media files. High-resolution media files, which are data-heavy, may display poorly on these devices, cause overheating, or drain the battery.

ANUKRITI KEDIA: A CASE STUDY ON DESIGNING FOR DISCONNECTION

In the markets we serve, most customers use low-end Android phones, juggle expensive data plans, and live in areas with limited network connectivity. This combination creates a fragile user journey—where even the most basic action, like checking an account balance, can feel frustratingly slow and unreliable. Yet for users of a fintech app, knowing their balance is often a prerequisite to conducting any transaction. If that single action fails, the entire experience collapses.

To address this, we adopted an offline-first design principle and united Design, Product, and Engineering around one shared mandate: to make resilience the default experience. Instead of assuming connectivity, we assumed disconnection. We began caching essential data like balances and user details directly on the device, enabling users to continue transacting even when the network failed. But just providing cached data wasn't enough; it was equally important to accompany the cached piece of information with a timestamp: "Last updated at 3:15 PM," a subtle but powerful gesture that helped us balance transparency with trust and control. They could instantly see how current their data was, transforming uncertainty into confidence.

We revisited our API architecture from the ground up. Redundant calls were eliminated, heavy responses compressed, and background refresh intervals intelligently spaced. What began as 20 API calls for a user flow was reduced to just 1, significantly reducing network and data dependency. From a user experience standpoint, we also tackled the invisible moments, the gaps between intention and confirmation. When data couldn't load, we replaced static spinners with contextual fallback states that clearly communicated what was happening. These small design details transformed waiting from a moment of doubt into a moment of assurance. This cut down our network dependency dramatically, reduced data consumption, and improved perceived loading time on low-bandwidth connections.

This journey taught us that Infrastructure Agnosticism is not simply about building for low bandwidth; it is about respecting unpredictability as a permanent design condition. In environments where connectivity, power, and cost fluctuate daily, resilience is paramount. Designing for these realities doesn't just make products technically efficient; it makes them emotionally dependable. Even when the world goes offline, the experience must remain online, steady, and human.

Researching in Limited Connectivity Contexts

For many of the reasons highlighted in this book and chapter, people with poor internet connectivity or low-end devices are often excluded from research, especially when we use remote or asynchronous research methods, such as unmoderated usability testing tools. As "AI interviewers" become more popular, the gap could widen further.

Our current methods typically involve participants with high-end devices and strong internet connections, so the experiences we design work well in those contexts but perform poorly in others. Many designers, engineers, and other product professionals also use iPhones or high-end Android phones and regularly have fast, reliable internet connections, further removing them from the user experience in limited connectivity contexts.

Here are some steps you can take immediately to get a wider perspective when designing.

Conduct Accessibility Audits

Accessibility audits are not new, but they are primarily used to adapt to assistive technologies, such as screen readers, or follow Web Content Accessibility Guidelines (WCAG).

The WCAG is a comprehensive toolkit for creating digital services and products that are accessible to a wide range of people with psychological and physiological disabilities. While some of the principles and recommendations apply to low-connectivity contexts (like Alt Text for images), WCAG does not fully capture the accessibility needs of this user group.

Just as physical disabilities create barriers to access, connectivity constraints and device limitations create equally significant barriers. As such, accessibility audits should also include the experiences of people with limited internet connections. This means applications and services should be tested in low-bandwidth and low-end device scenarios.

Given the varying speeds and coverage across Africa, testing should simulate different network conditions, including high latency and dropped connections. It should reflect real-world experiences by assessing load times, functionality, and data consumption in varying connectivity scenarios and speeds, including on 2G and 3G networks. Some key questions to answer: Does the app handle disconnections well? Can users still perform critical actions? Is data synced properly when the connection is reestablished? If content doesn't load properly, will the user or visitor miss important context? Do images load properly?

You can also track how much data different actions in your application or service consume to help you properly manage optimization efforts.

Various tools and techniques support this testing approach:

- Network throttling tools that allow developers to simulate various network conditions, including 2G, 3G, custom speeds, and latency
- Application Performance Monitoring tools that track metrics like load times and API response times in real time, which provide insights into how the service performs across different network types and device capabilities
- Crash reporting and analytics platforms that help spot issues related to low memory or network performance that plague lower-end devices and limited connectivity environments

Rethink Research Tools and Methods

Specialized research tools, such as usability testing platforms and diary study tools, have become increasingly popular in recent years. While these tools have greatly benefited research work, especially in remote settings, they require a reliable internet connection and consume substantial data, inadvertently excluding people with intermittent or poor internet connections, or those using low-end Android devices or feature phones.

As you plan your research, create sub-plans for different target audiences or connectivity scenarios. Have one plan for when technology works smoothly, and another for when connectivity is low or nonexistent. For example, if your prototype requires an internet connection to work, consider creating an offline version of screenshots or even printing out the screens for the prototype.

Collect User Insights

Simulations and audits can be helpful, but real-world data provides the best insights into actual user conditions. Use simple, data-efficient tools such as Google Forms or WhatsApp for diary studies and surveys rather than specialized, data-heavy apps. Going offline with, for example: paper diaries, contextual in-person interviews, or participatory design sessions, is also a viable, more inclusive alternative. We discuss these methods in more detail in Chapter 7.[17]

Observe when and how users interact with a service, paying particular attention to data management behaviors to understand the experience and identify opportunities for improvement. Beyond the quantitative data, collecting feedback on perceived speed and responsiveness can contextualize and support quantitative audits and tests.

Offer Relevant Incentives

In contexts of poor connectivity or data sensitivity, incentives go beyond being an added bonus to becoming a tool for inclusion and ensuring representative research.

Given that high data costs are a major challenge for a subset of African internet users, asking them to use their limited, expensive data for research activities could be limiting. Offering data packs as an incentive addresses this challenge, making participation more feasible and reducing this potential barrier to participation in remote or online research. For example, if users need to download a new app for a usability test or a diary study, providing data for the download and any associated usage makes it easier for them to participate in the research.

Key Takeaways

1. An offline-first design strategy is essential for inclusive design. By designing for the "worst case scenario" of slow or zero connectivity, you ensure reliability in every context
2. Users are often data-conscious and will abandon digital services that use too much data or do not provide data transparency
3. High network coverage does not equal usage due to cost and device limitations. Products must perform efficiently on low-end Android devices and 2G/3G networks
4. Standard remote research tools often exclude users with poor connectivity, so when designing research plans, ensure that tools and methods used for recruitment and conducting research do not exclude people in limited connectivity contexts

What Next?

Connectivity Audit

The goal is to assess how well your product works in various levels of internet connectivity.

1. Map out your user journey, assuming zero internet connection, to determine which features should still work in offline mode
2. Use network throttling tools to simulate 2G or 3G connection and attempt to complete the core tasks to identify where there are usability and access issues

Infrastructure Agnosticism Checklist

- Does the product fail gracefully as connectivity declines?
- Are all images and videos compressed for low-RAM entry-level devices?
- Are you avoiding rich media, complex animations, or other content that consumes excessive data?
- Have you deactivated autoplay and automatic background downloads?
- Do users have contextual feedback about connection state?
- Have you reduced redundant database roundtrips for basic info?
- Does the UI show file sizes before the user clicks "Download"?
- Can the user access basic functionality of the service even without an internet connection?
- Does the app or website work even on a slow connection? Is there a "low data" mode?
- Are there non-internet or physical fallbacks for essential tasks?
- Are you implementing lazy loading by prioritizing the most important content to load first?
- Are you tracking how much data different actions in your application or service consume?
- Are important details stored locally on devices to prevent redundant API calls?

Research Plan Audit

- Do you have backups in place if the internet connection fails during research?
- Are you using lightweight tools instead of data-heavy apps?
- Can people participate in research even with a poor internet connection?
- Are you offering relevant but noncoercive incentives, such as data packs to reduce participation barriers for cost-sensitive participants?

Notes

1. International Telecommunication Union (ITU). "Facts and Figures 2024: Internet Use." November 10, 2024. https://www.itu.int/itu-d/reports/statistics/2024/11/10/ff24-internet-use/
2. Matt Shanahan, "Despite Improvements, Sub-Saharan Africa Has the Widest Usage and Coverage Gaps Worldwide," *GSMA Mobile for Development* (blog). Last modified October 28, 2025. https://www.gsma.com/solutions-and-impact/connectivity-for-good/mobile-for-development/blog/despite-improvements-sub-saharan-africa-has-the-widest-usage-and-coverage-gaps-worldwide/
3. GSMA. "The State of Mobile Internet Connectivity 2025." Accessed August 30, 2025. https://www.gsma.com/somic/
4. Ookla, "Speedtest Global Index." Speedtest. Accessed September 1, 2025. https://www.speedtest.net/global-index
5. GSMA, *The Mobile Economy Sub-Saharan Africa 2024* (GSMA, 2024), 12. https://www.gsma.com/solutions-and-impact/connectivity-for-good/mobile-economy/wp-content/uploads/2024/11/GSMA_ME_SSA_2024_Web.pdf
6. GSMA, "The State of Mobile Internet Connectivity 2025." Accessed August 30, 2025. https://www.gsma.com/somic/
7. PowerSoft Africa, 2023 *Africa Domain Name Industry Study* (ICANN, 2024), 11. https://www.icann.org/en/system/files/files/africa-domain-name-industry-study-28may24-en.pdf
8. Best Broadband Deals, "The Cost of 1GB of Mobile Data in 237 Countries." Accessed August 30, 2025. https://bestbroadbanddeals.co.uk/mobiles/worldwide-data-pricing/#resources
9. Seacom, "Which Countries Charge the Most for the Internet in Africa?" February 16, 2022. https://seacom.co.ke/news/which-countries-charge-the-most-for-the-internet-in-africa
10. Laura Silver et al., "3. Mobile Owners: What They Struggle to Do," *Pew Research Center.* November 20, 2019. https://www.pewresearch.org/internet/2019/11/20/mobile-owners-what-they-struggle-to-do/
11. Steve Song, Jon Brewer, and Catherine Kyalo, *Free and Public WiFi Initiatives in Africa, Asia and the Pacific* (APC and Rhizomatica, 2024). https://www.apc.org/en/pubs/free-and-public-wifi-initiatives-africa-asia-and-pacific
12. City of Cape Town. "Public Internet Access in the City." Accessed August 28, 2025. https://www.capetown.gov.za/work%20and%20business/get-online/public-wifi-zones/public-wifi-across-the-city
13. World Bank Group. "Digital Transformation Drives Development in Africa." January 18, 2024. https://www.worldbank.org/en/results/2024/01/18/digital-transformation-drives-development-in-afe-afw-africa
14. International Telecommunication Union (ITU). "Facts and Figures 2024: Youth Internet Use." November 10, 2024. https://www.itu.int/itu-d/reports/statistics/2024/11/10/ff24-youth-internet-use/

15. Figures calculated from Simon Kemp, *Digital 2025: Nigeria* (DataReportal, 2025). https://datareportal.com/reports/digital-2025-nigeria; Kemp, *Digital 2025: Kenya* (DataReportal, 2025). https://datareportal.com/reports/digital-2025-kenya; Kemp, *Digital 2025: South Africa* (DataReportal, 2025). https://datareportal.com/reports/digital-2025-south-africa; and Kemp, *Digital 2025: Egypt* (DataReportal, 2025). https://datareportal.com/reports/digital-2025-egypt
16. As we have emphasized often, there is no one "African" experience. Africa is made up of 54 countries with different connectivity and access landscapes, even within countries. There is no blanket digital strategy for Africa, not just for the cultural reasons outlined in Chapter 5, but also because of the technical factors described in this chapter.
17. Chapter 7: The Unique Challenges and Solutions for User Research in Africa.

Bibliography

Best Broadband Deals. "The Cost of 1GB Of Mobile Data in 237 Countries." Accessed August 30, 2025. https://bestbroadbanddeals.co.uk/mobiles/worldwide-data-pricing/#resources.

City of Cape Town. "Public Internet Access in the City." Accessed August 28, 2025. https://www.capetown.gov.za/work%20and%20business/get-online/public-wifi-zones/public-wifi-across-the-city

GSMA. *The Mobile Economy Sub-Saharan Africa 2024*. GSMA, 2024. https://www.gsma.com/solutions-and-impact/connectivity-for-good/mobile-economy/wp-content/uploads/2024/11/GSMA_ME_SSA_2024_Web.pdf

GSMA. "The State of Mobile Internet Connectivity 2025". Accessed August 30, 2025. https://www.gsma.com/r/somic/

International Telecommunication Union (ITU). "Facts and Figures 2024: Internet Use." November 10, 2024. https://www.itu.int/itu-d/reports/statistics/2024/11/10/ff24-internet-use/

International Telecommunication Union (ITU). "Facts and Figures 2024: Youth Internet Use." November 10, 2024. https://www.itu.int/itu-d/reports/statistics/2024/11/10/ff24-youth-internet-use/

Kemp, Simon. *Digital 2025: Egypt*. DataReportal, 2025. https://datareportal.com/reports/digital-2025-egypt

Kemp, Simon. *Digital 2025: Kenya*. DataReportal, 2025. https://datareportal.com/reports/digital-2025-kenya

Kemp, Simon. *Digital 2025: Nigeria*. DataReportal, 2025. https://datareportal.com/reports/digital-2025-nigeria

Kemp, Simon. *Digital 2025: South Africa*. DataReportal, 2025. https://datareportal.com/reports/digital-2025-south-africa

Ookla, "Speedtest Global Index." Speedtest. Accessed September 1, 2025. https://www.speedtest.net/global-index

PowerSoft Africa. *2023 Africa Domain Name Industry Study.* ICANN, 2024. https://www.icann.org/en/system/files/files/africa-domain-name-industry-study-28may24-en.pdf

Seacom. "Which Countries Charge the Most for the Internet in Africa?" February 16, 2022. https://seacom.co.ke/news/which-countries-charge-the-most-for-the-internet-in-africa

Shanahan, Matt. "Despite Improvements, Sub-Saharan Africa Has the Widest Usage and Coverage Gaps Worldwide." *GSMA Mobile for Development* (blog). Last modified October 28, 2025. https://www.gsma.com/solutions-and-impact/connectivity-for-good/mobile-for-development/blog/despite-improvements-sub-saharan-africa-has-the-widest-usage-and-coverage-gaps-worldwide/

Silver, Laura, Emily A. Vogels, Mara Mordecai, Jeremiah Cha, Raea Rasmussen, and Lee Rainie. "3. Mobile Owners: What They Struggle to Do." *Pew Research Center.* November 20, 2019. https://www.pewresearch.org/internet/2019/11/20/mobile-owners-what-they-struggle-to-do/

Song, Steve, Jon Brewer, and Catherine Kyalo. *Free and Public WiFi Initiatives in Africa, Asia and the Pacific.* Association for Progressive Communications (APC) and Rhizomatica, 2024. https://www.apc.org/en/pubs/free-and-public-wifi-initiatives-africa-asia-and-pacific

World Bank Group. "Digital Transformation Drives Development in Africa." January 18, 2024. https://www.worldbank.org/en/results/2024/01/18/digital-transformation-drives-development-in-afe-afw-africa

Chapter 5

Building for Sociocultural Context

A few years ago, IM Mastery Academy launched a campaign in Johannesburg, a multicultural market. As Tochy Emereole, a marketing professional, puts it: "One of the key headlines included a word that, while harmless in English, carried a negative cultural connotation for certain South African communities." They had to pull back the campaign and rework the copy in collaboration with local linguists and cultural consultants to identify terminology that was universally positive across Zulu-, Xhosa-, Afrikaans-, and English-speaking audiences. They also had to change the imagery to make it more representative of the diverse community. When they ran their post-campaign survey, they found that their trust score in South Africa was 34% higher than in other regions where the campaigns were not localized.

This story demonstrates the importance of cultural understanding for success in any market.

Understanding Cultural Cognition

Culture is broadly defined as information shared among groups of individuals. Around the world, we see variations in different aspects of cultures, including practices, norms, languages, and

DOI: 10.1201/9781003648192-7

styles of social interaction. These variations often arise from environmental and historical ways of being and become entrenched in cultural norms.

These cultural practices could have an impact on cognition, shaping the way people think, reason, categorize, make decisions, form expectations, and perceive the world. What is normal or expected, the meanings assigned to different images and symbols, are all influenced by culture. This phenomenon is known as cultural cognition—how what we think and perceive is shaped by our sociocultural context.

Navigating cultural differences when offering digital services outside your country requires cultural intelligence. Coined by Earley and Ang,[1] Cultural Intelligence refers to understanding, respecting, and adapting to cultural nuances—a capability necessary for global competitiveness and business success. It is about being able to identify cultural differences and adjust behavior and

FIGURE 5.1 Even words can mean different things.

decisions accordingly. Cultural intelligence begins with "Cultural Drive": the willingness to learn about and engage with other cultures.

This chapter highlights how cultural nuances influence user behavior and preferences and discusses how to incorporate cultural sensitivity into design, product, marketing, copy, and even business decisions. The individual's user experience and expectations of your product, how they interact with it, what they use it for, whether they think they can use it, how they respond to it, and so much more are shaped by their sociocultural context. The more your product resonates with and accommodates cultural nuances, the likelier it is to succeed. Many businesses have failed to grow because they simply copied and pasted what had worked in another country without considering the unique cultural context.

This applies to different kinds of products and services, from global retailers to international movie launches, from restaurant and fast-food chains to software, and from UX writing to marketing.

Throughout the chapter, we dive into specific design considerations to show how cultural awareness can lead to more effective and successful products.

Cultural Classification Models

Various cultural classification models provide a useful lens for understanding how culture shapes behavior and expectations. These frameworks provide a high-level view into the psychological and sociological concepts and cultural values that underpin behavior and have useful applications for designing interactions and experiences in African contexts.

Many models exist, including Hofstede's Cultural Dimensions Theory, Hall's High-Context versus Low-Context Cultures, Trompenaars' Seven Dimensions, the GLOBE Cultural Dimensions, and the Lewis Model of Culture. Though many of them were developed in the context of leadership and organizational behavior, they provide valuable insights for product development and design. This chapter explores the three most prominent models.

It is important to note that these frameworks are not about what is good or bad, nor do they go into detail on the reasons for each culture's norms. They are simply a broad descriptive and classification tool, based on research at specific points and in specific contexts, of diverse cultures and their practices and behaviors. Furthermore, they should be seen as guidelines rather than fixed rules and as a starting point for further primary research to avoid stereotypes.

Hofstede's Cultural Dimensions

One of the most widely known frameworks for understanding cultural differences is Hofstede's Cultural Dimensions Theory, developed by Dutch cross-cultural psychologist Geert Hofstede. It was based on research he conducted with a sample of IBM employees across the globe between the 1960s and 1970s.

The framework explores the connection between cultural values and behavior, which can be extended to expectations and perceptions. It comprises six dimensions: collectivism-individualism, power distance, masculinity-femininity, uncertainty avoidance, long-term versus short-term orientation, and indulgence-restraint. These dimensions exist on a spectrum, rather than being binary, allowing for greater nuance in understanding these cultural differences.

While an in-depth discussion of this framework is beyond the scope of this chapter and book, here is a high-level overview.

- Collectivism-Individualism measures the extent to which people see themselves as part of a larger group. In cultures closer to the individualism side of the spectrum, personal goals are prioritized over group goals, and self-reliance is highly valued
- Power Distance Index (PDI) measures the extent to which less powerful members of society accept and expect unequal distribution of power. Countries (or cultures exist on a spectrum from High PDI to Low PDI. In High PDI cultures, titles, status, and other such symbols of hierarchy are important and accepted, and authority is not to be questioned. On the other end of the spectrum, there

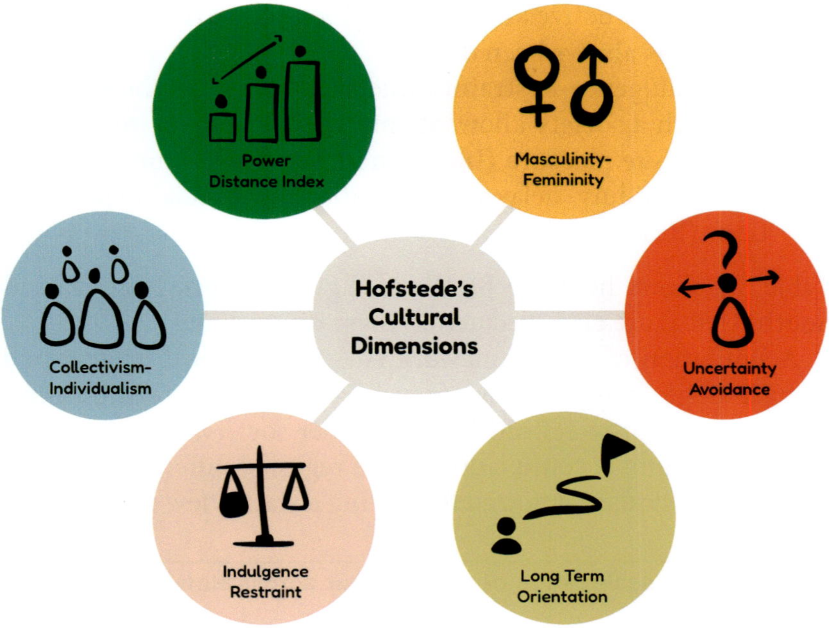

FIGURE 5.2 Hofstede's cultural dimensions.

is a greater focus on equal power distribution and flat structures

- The Masculinity-Femininity dimension is based on expectations of emotional gender roles. Masculine-dominant cultures value competition and material success over caring for others and cooperation, characteristics of feminine-dominant cultures
- Uncertainty Avoidance relates to how a group reacts in the face of uncertainty. High Uncertainty Avoidance (HUA) cultures place greater importance on rules, predictability, and formality, while Low Uncertainty Avoidance (LUA) cultures value flexibility and informal structures. LUA societies are more comfortable with ambiguity and change than HUA societies
- Long-term Orientation cultures are future-focused and always preparing for an ever-changing world with adaptability, while Short-term Orientation cultures view the

past as instructive and are more focused on tradition and "what has always been done"

- The Indulgence-Restraint dimension measures the extent to which a society allows its members to follow impulses and pursue pleasure (Indulgence) versus the importance placed on duty, delayed gratification, and social expectations (Restraint)

Table 5.1 shows how the Big Four (Egypt, Kenya, Nigeria, and South Africa) are classified across these dimensions.

A limitation of this framework is that it is based on research from a single company and focused on business relationships. Thus, it may not be reflective of the wider society, especially as it doesn't take into account cultural and regional differences within a country. It is also criticized for having a static view of cultures.

TABLE 5.1 Classification of the Big Four across Hofstede's Dimensions

Dimension	Egypt[a,b]	Kenya[a,b]	Nigeria[a,b]	South Africa[c]
Collectivism-Individualism	Collectivism	Collectivism	Collectivism	Collectivism
Power Distance Index	High	High	High	Mid
Masculinity-Femininity	Masculine	Masculine	Mid	Masculine
Uncertainty Avoidance	Mid	Mid	Mid	Mid
Long-term Orientation	STO	STO	STO	STO
Indulgence - Restraint	Restraint	–	Indulgence	Indulgence

[a] The Culture Factor Group. "Country Comparison Tool." Accessed August 28, 2025. https://www.theculturefactor.com/country-comparison-tool

[b] Geert Hofstede, "The 6-D Model of National Culture." Accessed August 20, 2025. https://geerthofstede.com/culture-geert-hofstede-gert-jan-hofstede/6d-model-of-national-culture/

[c] Pedro Henrique D. Pecly and Priscilla Cristina C. Ribeiro, "Entrepreneurship in the BRICS and Cultural Dimensions," *Brazilian Journal of Operations & Production Management* 17, no. 2 (2020): 1–18. https://doi.org/10.14488/BJOPM.2020.013

Edward T. Hall's High-Context and Low-Context Cultures

Hall's classification of cultures as high-context or low-context[2] describes how cultures communicate and is considered a foundational model for understanding cultures.

In high-context cultures (HCCs), people rely heavily on additional cues beyond content—body language, gestures, the relationship between speakers, the time and place of communication, and the use of silence—to interpret what has been said (or left unsaid). Messages are often implicit, and communication is indirect. People are expected to "read between the lines." HCCs also prioritize personal relationships, societal hierarchies, and tradition.

In low-context cultures (LCCs), on the other hand, communication is direct and explicit. Meanings are clearly stated, and asking clarifying questions is not only welcome but expected as a sign of active listening—whereas in high-context cultures, asking clarifying questions could be perceived as rude. LCCs prioritize individual accomplishments over group achievements, favor decentralized authority, and value individuality and privacy.

In *Beyond Culture*, Hall notes that "no culture exists exclusively at one end of the scale."[3] However, they tend toward one orientation and can be arranged on a continuum from extremely low-context to high-context. A limitation of this model is that it focuses solely on one dimension of communication, a small part of any culture and language, and ignores other aspects that can influence communication styles, such as individual personality, which he acknowledges in his book *The Silent Language*.[4]

Although his framework doesn't explicitly mention African cultures and nations, the descriptions given can be applied to these cultures, and most of them are classified as high context.

Trompenaars Hampden-Turner 7 Cultural Dimensions

Developed in the early '90s by Trompenaars, this framework covers seven dimensions of culture that examine how people navigate relationships, time, and environment. It is based on research across 40 countries and is the most recent of the three. Their work is based on the idea that people raised in the same culture learn similar values and norms, which become ingrained in their psyche.

Trompenaars' research suggests that cultural shifts can be more dynamic than Hofstede's view allows. These cultural shifts can be because of globalization, generational differences, technology, and migration. Urban areas are most impacted by these changes.

The seven dimensions include:

- Universalism versus Particularism: This is the cornerstone of this model and is about whether or not a culture applies rules consistently. In Particularist cultures, rules are applied based on specific circumstances or relationships, while Universalist cultures prioritize rules and fairness
- Individualism versus Communitarianism: Individualist cultures focus on the self, while communitarianist cultures focus on common goals and objectives. This is similar to Hofstede's individualism-collectivism dimension
- Neutral versus Emotional: This dimension refers to whether emotions are openly expressed or not. In neutral cultures, emotions are held in check, while in emotional cultures, feelings are expressed openly and naturally
- Achievement versus Ascription: This is about status and how it is accorded to people. Achievement-oriented societies prioritize high achievers, while ascription-oriented societies base status and power on titles, age, family background, and similar factors. This is somewhat similar to Hofstede's Power Distance dimension
- Internal control versus External control: This refers to the extent to which people believe they have control over their environment. Internally directed cultures focus on how they can control their external environment to achieve goals, while externally directed cultures conform to the external environment
- Sequential versus Synchronic: This dimension is about how a culture approaches time. Sequential cultures often focus on one thing at a time and view time as a finite resource that can be wasted, whereas synchronic cultures are often more flexible, willing to multitask and engage in multiple projects simultaneously

- Specific versus Diffuse: This dimension is about how much one's personal and professional lives are expected to be separate. In Specific cultures, there are clear boundaries between work and personal life, while in Diffuse cultures, these spheres can overlap, and relationships can blend both personal and professional interactions. Trompenaar describes this dimension as similar to Hall's high-context/low-context model

A criticism of this framework is that its dimensions are binary, whereas in reality, cultures may exhibit both ends of a dimension depending on the context.

Applying Frameworks Responsibly

Overrelying on a single framework or treating it as a strict guideline can lead to stereotyping. This results in product design that feels inappropriate or generic, potentially sparking backlash, which can result in churn, poor brand reputation, and low or no adoption.

The contradictions between the frameworks also highlight the importance of not just sticking to one. While overlaps exist—for instance, Hofstede classifies Egypt as collectivist and Trompenaars' research revealed Egyptian respondents as communitarian, showing alignment between the two frameworks on this dimension—the frameworks also differ in how they classify countries. While Hofstede classifies Nigeria as highly collectivist, in Trompenaars' research,[5] Nigerians were among the highest-scoring individualists based on the respondents in his survey.[6] Respondents from Kenya, which might be described as high context by Hall's framework, largely disagreed with the statement that respect depends on family background, which is common in high context cultures.

As Erin Meyer notes in her book *The Culture Map*,[7] cultural scales and dimensions are most effective when applied with an understanding of cultural relativity—that is, cultural behavior is not about absolute standards but about the relative position of one culture to another on a specific scale. In addition, it is important to recognize that within each culture, there is a range of what is considered acceptable or not acceptable.

THE COLLECTIVIST MYTH

Many founders and builders assume that community-driven initiatives will always work in African contexts, but that is not always the case.

Research done by YUX Design among young women in Ghana showed that this population group was increasingly avoiding collective savings groups (esusu). They still want to save, but they don't want anyone else in the community to know about their projects or their money.

This echoes insights from research Lade did in Kenya and Nigeria on financial behaviors and attitudes regarding agents and collective savings, respectively. Even when people want to participate in these communal savings, they still have a strong desire for privacy. Participants in the Nigeria esusu study specifically requested that only the group admin should be able to see who has paid and who is owing. People wanted to avoid being shamed by others in the group for owing.

Some participants in the Kenya study mentioned preferring ATMs over agents to avoid social interactions they didn't want. They were more concerned about the privacy of their transactions.

These examples show that culture evolves and that while these frameworks may provide a good starting point, primary research is essential to gain a more robust and accurate perspective and avoid the potential problem of stereotyping cultures. It also highlights that culture is not limited to a country, but can be a multifaceted, complex web of regions, gender, age, ethnic groups, or even socioeconomic class.

Localizing Experiences Based on Cultural Frameworks

Many African countries tend toward high-context communication and score higher on collectivism, power distance, and uncertainty avoidance. These different dimensions could have a great impact on user preferences—how information should be presented, how much control users expect over their experience, and what users expect from a product. The implications extend to product offerings, branding and advertising, communication styles, and more.

For example, collectivist cultures may value collaborative features over personalized ones or may be interested in seeing what others think when deciding how to use a feature. Where uncertainty avoidance is high, users may prefer predictable interfaces and experiences or may hesitate to explore different parts of the solution for fear of "breaking something."

UNCERTAINTY AVOIDANCE IN THE FIELD

When Lade worked on a project for users new to smartphones, her research showed that they ignored unfamiliar concepts during onboarding. This means that they missed out on essential information. Because phones are expensive, especially compared to income, they were afraid of breaking or damaging something, making them unlikely to explore.

Design can cater to these users by providing more guidance and explanations in the interface and during onboarding, limiting hidden gestures, and emphasizing benefits to users. The myth of "good design doesn't need to be explained" definitely does not apply here. While conducting usability studies with new-to-the-internet participants, we observed that some were hesitant to explore the apps they were invited to test. Many said they'd just call someone for help, and if they didn't find help, they would do nothing, abandoning the app.

If your target audience is higher on the power distance continuum, they may prefer authoritative visual elements and messaging that positions the service as the expert. In cultures that are specific rather than diffuse, people might have multiple social media accounts for work/career and various personal interests, since these are expected to be kept separate.

Understanding these cultural norms and differences provides a foundation for making strategic decisions to localize user experiences. This ensures usability and usefulness, which in turn impacts engagement, retention, and the long-term success of digital products. However, this does not necessarily mean you have to design new and unique experiences for each country or target audience. You can offer similar experiences to countries with similar cultural dimensions and practices. For example, based on

Hofstede's dimensions, Nigeria, Egypt, Kenya, and South Africa all score mid-low on Uncertainty Avoidance.

From Awareness to Action

Cultural Intelligence, as described above, goes beyond awareness and knowledge. It also involves developing culturally sensitive strategies based on an understanding of the culture—what Earley and Ang refer to as "cultural strategy." Cultural action then follows: implementing these strategies for culturally appropriate behavior. In the context of product design, this means creating culturally appropriate products and making culturally appropriate design and copy decisions. In this section, we'll explore how cultural intelligence translates into thoughtful design decisions.

Culturally intelligent teams adapt products to the cultural realities/nuances of specific markets through localization, leading to international success. When people talk about localization, they often mean "offer the service in the local language." But localization, like culture, goes beyond language and translations to encompass sociocultural awareness. True localization, which we call **hyper-local customization**, is really about accessibility; it means culturally reframing an offering to fit the local context. This involves customizing features (by adding or removing them) and ensuring appropriate positioning and marketing. Key elements to consider include communication styles and language, imagery and symbolism, and environmental context.

Language, Communication Styles, and Names

Language is a huge part of culture. It is a system of communication that allows the expression of thoughts, feelings, and ideas, which can be spoken, written, or signed. Language goes beyond grammar and syntax to include slang, social norms, and cultural information that add meaning to what is being said.

Translations

Localization begins with translation: offering users the option to access a service in the language they're most comfortable with.

Translation improves accessibility, understanding, and overall experience.

However, you should be careful about default language settings because location does not equal language understanding or familiarity. Due to colonial histories across many African countries, English, French, and German could be part of the lingua franca alongside local languages that people communicate more readily with. South Africa, for example, has at least 10 official languages alongside dozens of other recognized languages. Furthermore, "international" languages have localized versions. For example, there is Nigerian English, Benin Republic French, Egyptian Arabic, and so on.

While people typically prefer to engage with content in their own language,[8] they may prefer another language. Take, for example, a Nigerian relocating to Kenya who prefers to access sites in Swahili to aid language learning. This use case is common for many Africans for whom English is not their first language or official language, yet who prefer it for aspirational purposes. Ultimately, users should have various language options to choose from.

EXPERT INSIGHT: YUX DESIGN ON LANGUAGE

When testing products for users in Tanzania and Rwanda, the YUX team found that they preferred to interact with the interface in English rather than Kiswahili. This behavior was driven by two key motivators:

- Educational Utility: Users view the digital interface as a language-learning tool. By navigating an app in English, they are practicing a skill they find valuable for their future
- Social Signaling: Mastering a "foreign" or "international" language is often perceived as a sign of higher education and social status

However, offering multiple languages is only effective if translations are culturally appropriate. Relying solely on literal or AI-generated translations may not cover all the nuances of language—humor and idiomatic expressions in particular could be lost in translation—and so it is better to work with expert translators and test copy and translations with native speakers.

KOME SIDESO: A CASE STUDY ON FAILED TRANSLATIONS

In a project developed for users in Angola, the team initially attempted to use a digital translation tool. The automated tool translated technical and cultural terms too literally, making the product unappealing and difficult to understand. The team realized that in Lusophone and Francophone markets, if the language isn't handled with cultural nuance, "nobody is going to listen."

They moved away from automation and hired local translators to provide contextually accurate translations. The impact was immediate: they saw an increase in app downloads once the language felt authentic.

Multilingualism

Many Africans are multilingual. Beyond "standardized" international languages and their localized versions, people often speak multiple local languages—and frequently mix them within a single interaction. This practice, known as code-switching, is a common communication style across the continent, appearing in face-to-face conversations, online platforms, and voice-based messaging. For example, a South African user might begin a tweet in English and seamlessly transition to a local language midway.

Code-switching presents an interesting pattern to design for. Take voice agents, for example. If users are more comfortable expressing themselves in multiple languages and the agent does not understand them, that results in a poor experience. Research Lade did on voice interactions showed that participants spoke with voice assistants the same way they would speak with people. They spoke in multiple languages, transitioning between English and another language. A user might start a request in English and finish it in Yoruba or Pidgin. Current voice agents often fail here because they are trained to listen for a single language at a time.[9]

Another consideration for multilingual users is that they prefer different languages in different contexts. A user in Kenya might prefer to type in English but speak in Swahili.

One way to accommodate multilingualism is to let users set the language they want to use to interact with the system. For voice and chat-based interactions in particular, tools should be built with datasets that represent a wide range of accents, dialects, and languages, and can grasp user intent even when they switch language midway.

Cultural Terminology

Another important consideration is consistency with local expectations regarding terms, names, currencies, time zones, slang, and other related conventions. A common example is "chips" in American English versus "crisps" in British English. Another example is the word "stove." In Western contexts, most people imagine a gas or electric stove top, but a Nigerian would think of a portable kerosene camping stove for cooking. What the average Dutch or American would call a stove, a Nigerian would call a "cooker," which conjures different images: pressure cooker, slow cooker, rice cooker, multi-cooker, and so on. This is an example of cultural information adding meaning.

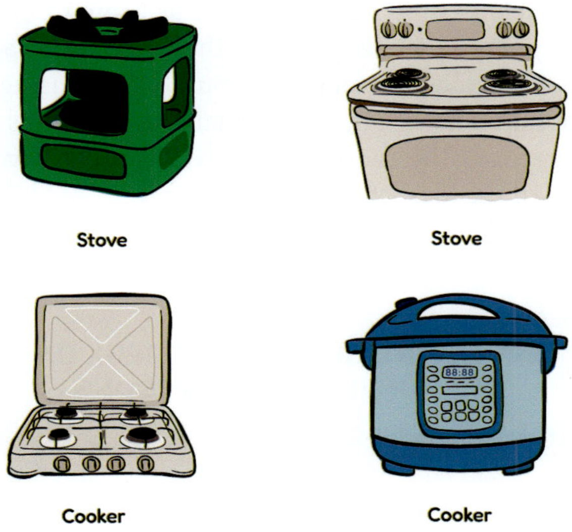

FIGURE 5.3 What does a stove or cooker mean?

Imagery and Symbolism

Imagery and symbolism are deeply ingrained in culture and function as powerful communicators of meaning, values, and identity. In design, they move beyond aesthetics to influence how users perceive and interpret digital interfaces and experiences.

Design elements—icons, color associations, symbols, and visual aesthetics—are non-universal in both existence and associated meanings. The same element can mean different things in different cultures.

Understanding the non-universality of these elements is essential for designing products that resonate positively with diverse global audiences. Overlooking or misusing them could lead to backlash.

Iconography

A common assumption in design circles is that everyone understands icons to mean the same thing. But this is not true. People interpret icons differently based on their experience with technology, cultural background, location, accessibility needs, or other existing knowledge.

Because of this reality, treating icons as universally understood is an ineffective design practice. Depending on who is looking at it, the search icon could be interpreted as a ping pong paddle[10] and scuba flippers as a hijab.[11] Using the dollar sign ($) to represent currency, a wallet, or spending could be confusing to people who are not familiar with it, as their daily transactions occur in a different currency. The more appropriate approach would be to use a symbol of the local currency: KSh for Kenyan Shillings, ₦ for Nigerian Naira, R for South African Rand, and so on.

EXPERT INSIGHT: YANN LE BEUX ON NON-UNIVERSALITY OF ICONS

Icons don't always speed things up; they can actually be time wasters if users spend time analyzing them. Field research with farmers in Zambia and vendors across West Africa reveals that users often spend significant amounts of time trying to decipher icons and uncover their meaning. Many "standard" icons refer to objects that no longer exist in the user's physical world or may have no cultural relevance. Symbols like the floppy disk for "saving" or the vintage microphone for "voice notes" can be meaningless. Instead of a generic bus icon, the Kenyan version of an app might use a "matatu." Simply redesigning key icons to match local visual mental models can dramatically improve user confidence and understanding.

Testing icons with participants of diverse backgrounds is essential. This can be done via usability testing, recognition, word association, and card sorting. Also, icons should always be accompanied by text labels, even if they seem widely known or self-explanatory. Labeling icons may feel redundant to designers, but it ensures comprehension for users. Another solution is using adaptive iconography: changing the icon based on the country. This design pattern is commonly used for currencies.

Color

Beyond iconography, the meanings of colors and other symbols may differ across cultures. A color that signifies trust and reliability in one culture might represent danger or mourning in another. Red, for example, is associated with good luck in some Asian cultures, anger or warning in many Western cultures, and danger or mourning in some African cultures. White commonly connotes peace and purity (positively coded emotions), but in some Asian and African cultures, it is associated with mourning (negatively coded emotions). Understanding these nuances is a first step in ensuring that visual aesthetics resonate positively with the target audience.

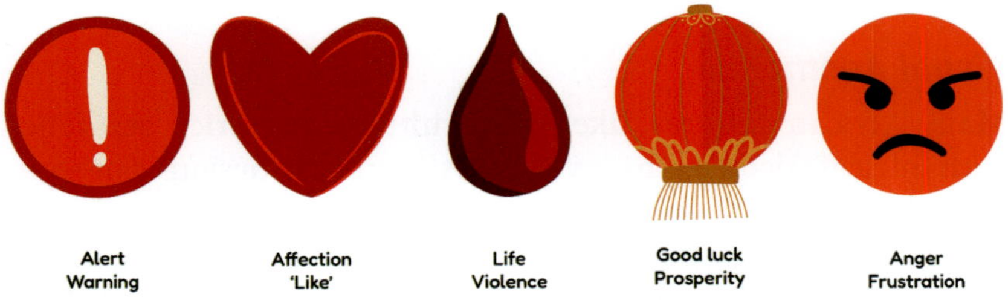

FIGURE 5.4 The color "Red" in different contexts.

Within Africa alone, color associations may vary by region, as the continent is not a monolith. For example, black is generally seen as signifying death; however, in Egypt, it may symbolize rebirth and is a celebratory color among the Idomas in Nigeria.[12] The point is not to entirely avoid certain colors but to understand their potential meanings and interpretations and use them thoughtfully. However, it is important to note that people may not always have such strong associations with the meanings of colors from their culture.

Symbolism

When symbols are thoughtfully integrated into product design, they demonstrate cultural understanding and create a sense of familiarity among users, which can enhance the overall user experience. However, symbols, such as animals and other motifs and elements, should be carefully assessed before use.

Animals that are revered or neutral in one sociocultural context may be viewed negatively in another. Owls, for example, connote wisdom in Western culture but are seen as evil or bad omens[13] in some Kenyan and Zimbabwean cultures, and are perceived as stupid in some Indian cultures.

Designers can integrate cultural motifs and patterns into digital interfaces, onboarding designs, stickers, emojis, and more. Examples include textile designs across Africa: the Kente of various Ghanaian groups, the Shuka of the Maasai in Kenya, the Ndop of the Bamileke, and the Tibeb/Tileb patterns of Ethiopia. The sense of familiarity that these symbols and motifs evoke could improve a user's willingness to engage. But designers must be careful not to veer into stereotypical or excessive use of these motifs. Testing via participatory design methods can help ensure symbolism remains authentic and appropriate.

Visual Aesthetics

Many African cultures, like Asian cultures, have rich aesthetics with intricate details. These visually stimulating designs can seem overwhelming to those accustomed to Western minimalism—the "clean" aesthetic popularized by Apple with its extensive use of white space. However, minimalist design may be out of place in African contexts, where users expect images, prints, and the like.

There is another layer to this: when you combine an expectation of vibrancy and dense aesthetics with issues like poor connectivity,[14] excessive white space can seem like missing information or incomplete loading.

Over multiple research projects with participants in Nigeria, Lade observed that when participants encountered a lot of white space on in-app interfaces, they became confused. They would refresh the page over and over again. This behavior was more prevalent the newer they were to the internet. When probed, they revealed that they thought the page had not loaded completely. They were used to vibrant and dense experiences in their offline lives. Furthermore, most of the time they encountered white space online, it was because the internet connection was poor and the page had failed to render.

ANUKRITI KEDIA: A CASE STUDY ON REDEFINING SIMPLICITY FROM MINIMALISM TO CLARITY

Like many design teams influenced by global standards, Anukriti's team began with the assumption that Western ideals of minimalism—streamlined interfaces, deep navigation hierarchies, and tightly grouped information—would translate seamlessly to African contexts. They equated simplicity with usability and elegance with effectiveness.

In practice, the opposite was true. Through usability testing and live feedback, they quickly discovered that many users felt lost inside the app, often unable to locate core functions or even realize they existed in the first place. For users new to digital financial tools, abstraction didn't reduce cognitive load; it increased it. They didn't want to explore or guess. They wanted clarity. They wanted to see everything they could do without worrying about what might be hidden.

This realization led the team to rethink their entire approach. They flattened the navigation hierarchy, brought key functions to the surface, and made high-frequency actions immediately visible. They paired icons with clear text labels, creating affordances that worked across literacy levels and languages.

The team stopped designing for the idealized "global" user and started designing for the lived reality of their own. Their goal shifted from aesthetically minimal to cognitively minimal—reducing friction not through fewer elements, but through clearer pathways.

Usability isn't universal. Patterns that work elsewhere can fail spectacularly here because the user's digital context is different. Many are multitasking, using shared devices, or operating under pressure and uncertainty. In those conditions, the most respectful interface is one that's explicit, transparent, and easy to trust. Sometimes, that means replacing minimalist purity with functional maximalism.

Designers can address cultural diversity by creating adaptable design systems where components, templates, and principles can be adjusted based on cultural context. For example, using different images in different locations or layouts that work for both right-to-left and left-to-right languages.

Social and Environmental Context (Interaction Patterns)

Understanding the environmental context in which your target audience lives goes a long way in ensuring a successful product with a seamless user experience. For example, a newly launched app in Nigeria that accepts only credit cards as a payment method would not be widely adopted because credit cards are not common in most sub-Saharan countries.[15] Adoption can be achieved or improved by using locally relevant payment methods, such as mobile money in Ghana, pay by bank transfers in Nigeria, and M-Pesa in Kenya. Netflix is a good example of this because it adopted local content and created new pricing plans that fit into the context of the intended customer audience.

A NETFLIX CASE STUDY

Netflix officially launched in Africa in January 2016, focusing on South Africa, Nigeria, and Kenya as part of its global expansion.

First, it acquired more locally relevant content and invested in original productions to improve its content offering. Then in 2021, following successful launches in Asian countries, it introduced a mobile-only plan across African countries, including Egypt, to grow its user base. This plan allowed users to access Netflix with just their Android device and a valid email address. While Netflix doesn't release its subscriber numbers, estimates suggest the mobile-only plan drove a 50% increase in India within a year.[16] Similar growth likely occurred across African markets. Netflix eventually started charging for the mobile-only subscription, but it is still priced lower than other subscriptions on offer.

Another example is designing for shared device usage. Shared devices are quite common in many African countries, so products should be designed in that context. One approach is to enable multiple profiles with adequate security and privacy measures.[17]

Address Formats

Address formats vary across countries. In some, postcodes are required when filling out forms; in others, they are not. In Nigeria, for example, including postcodes in address data is not common, and most people don't even know their postal code. Across the continent, people sometimes rely on landmarks to navigate, even when official addresses exist. Designing flexible address fields that accommodate non-standard formats is one way to address this, for example, including a "Landmark" field.

Names and Naming Conventions

African names tend to be long and include symbols. However, forms typically impose character length and font limits on name fields, which can create a poor experience for users who cannot enter their names correctly or completely. This can be particularly problematic in high-stakes contexts like financial services, travel, and healthcare.

Furthermore, name ordering varies across contexts. In some countries, first, middle, and last names are required, whereas in others, only one may matter, and the first name may come after the last name. Kome Sideso, Deputy Head, CcHUB Design Lab, says that in Rwandan naming systems, last names may not be as important as first and middle names, which are usually sufficient for forms.

As with address fields, name fields should also be flexible to allow users to input their names correctly regardless of length and characters.

AI tools with voice interfaces also require particular attention. Voice models should be trained on diverse multilingual datasets—including local names, pronunciations, and language variations—to ensure accurate name pronunciations.

Multiple Phone Numbers

In many sub-Saharan countries and other developing economies, people often have more than one SIM to separate business and personal life, take advantage of discounts and promos, and have backups during poor connectivity, among other reasons. Until recently, users in this group resorted to workarounds like creating WhatsApp Business accounts (even without having a business), downloading cracked versions of WhatsApp, or using third-party apps created to meet this need. This exposed them to possible security risks. Now WhatsApp has adapted to this sociocultural need by allowing up to two accounts on the same device.

THE FRAGMENTATION OF DIGITAL CULTURE: WE'RE ALL DIFFERENT, EVEN IN THE SAME COUNTRY

While launching a pilot in Kenya, Princess Akari found that despite strong product-market fit in Nigeria, the core value proposition failed because M-PESA had already solved the pain points that existed in Nigeria. Similarly, Olabinjo Adeniran noted that even neighboring countries like Ghana and Nigeria developed diverging user behaviors based on their unique environments.

Behavior can be shaped by infrastructure and access, which then coalesces into culture. Adeniran observed that in Ghana, widely available public Wi-Fi and telco-provided "free social media bundles" created a culture that consumed video content voraciously. In contrast, high data costs in Nigeria kept users in a "text-and-image-first" mode for longer. These infrastructural differences created distinct digital cultures with different expectations of what makes a product "usable." A feature that feels essential in a high-bandwidth culture might feel like a data drain in a low-bandwidth one.

Likewise, from his growth work, Oluwapelumi Oyetimein discovered that effective communication requires granular localization, even within the same country. Scaling from Lagos Island to the Mainland required shifting tone; moving to Ibadan demanded the use of Yoruba and Pidgin. Expanding to Northern Nigeria meant completely rethinking gender dynamics in marketing and cultural references.

To succeed, builders must move beyond the idea of "one African product" and build localization infrastructure that adapts to the specific country and community, rather than the continent as a whole.

Researching for Culturally Sensitive Design

The cultural contexts and frameworks explored in this chapter are by no means an exhaustive list, but only a starting point for considerations in your business, product, and design choices as you look to explore an international audience. You will need to identify and consider additional nuances. Although these frameworks may be generally applicable, specific cultural dimensions can differ across groups within the same country and change over time. The most effective approach would be to conduct your own on-the-ground research.[18] Keep the following in mind when researching cultural nuances.

- Understand the Market: Look at demographics, the political landscape, economic outlook, competitive analysis on local competitors, and other reports from reputable research organizations and agencies. These give a high-level view of important trends that can help provide foundational context

- Understand the Current Cultural Landscape: Look into the sociocultural landscape, including popular design trends, color and symbolism, communication styles, cultural terminology and slang, and cultural attitudes towards technology. Like examples we have covered in this chapter, this can also give you an overview of the expectations of the potential users and a starting point for your primary research
- Conduct Ethnographic Research: Beyond secondary research, immersive, on-the-ground experiences in target countries can provide the richest insights. As you observe and engage with potential users in their real-life context, you can better understand the user journey and experience outside of the context of the product. Direct engagement uncovers subtle behaviors, unspoken norms, and emotional responses that may not be captured in frameworks, surveys, and generalized reports. This approach helps you avoid the trap of leaning too much on stereotypes, a major criticism of fixed frameworks and categorizations of cultures
- Collaborate with Local UX Professionals and Cultural Experts: Whether or not you'll be physically present, work with researchers, designers, and other experts who understand the local context, either because they are native or regularly conduct research with the target audience. This is especially important because of the bias in which respondents may change their behavior during a study—a phenomenon known as the researcher effect—and this can be compounded when the researcher is of a different ethnicity or culture. Local experts can provide authentic insights, help navigate cultural sensitivities, and ensure that research questions and artefacts are culturally appropriate and participant responses and behaviors are accurately interpreted
- Use Inclusive Recruitment and Research Methods: Being physically present can be expensive. If you're on a smaller budget, rethink your tools and participant

recruitment strategies. You can find people in your own city who may have similar experiences or come from the sociocultural background you're interested in, as a starting point or for a "sanity check." For example, if one key sociocultural factor to consider is poor or limited internet connectivity,[19] then you can find people who experience this issue. Instead of using specialized tools for research (e.g., diary studies), you could use simple online forms or commonly used instant messaging tools like WhatsApp

Key Takeaways

1. Shift from translation to hyper-localization, that is, going beyond simple translations to working with native speakers and designing for multilingual speakers, especially in the context of voice interactions
2. Design for cultural cognition, understanding that imagery and symbolism are not universal and are interpreted through their cultural lens
3. Design for real-world behaviors, accommodating environmental realities such as how addresses and names are written and described
4. Avoid stereotypes in applying cultural frameworks by conducting primary research

What Next?

Cultural Sensitivity Audit

Language and Communication Style

- Have all core UI elements been reviewed by native speakers or professional translators to ensure linguistic nuance is covered?
- Have you given users the option to choose what language they want to use the service in?
- Can the system recognize multilingual user inputs?

Imagery and Symbolism

- Have you localized icons and symbols to what is culturally relevant?
- Have you tested the understandability and suitability of icons and symbols with participants?
- Are icons clearly labelled to aid understanding?
- Have you tested your color palette for regional meanings?
- Are visual patterns and motifs appropriately used?
- Are imagery and photos representative of the users?
- Do you have an adaptable design system that can be adjusted based on cultural context?

Social and Environmental Interaction Patterns

- Are address forms flexible to support local ways of writing and describing addresses?
- Can the name fields accommodate short or long names and varying order requirements?

Research

- Have you done research to understand the sociodemography of the target audience?
- Do you understand the cultural landscape?
- Have you conducted your own primary research with users from the region you're interested in?
- Are you collaborating with local UX professionals and cultural experts?
- Are you refining your research and recruitment methods to be inclusive?

Notes

1. P. Christopher Earley and Soon Ang, *Cultural Intelligence: Individual Interactions Across Cultures* (Stanford University Press, 2003).
2. Context refers to the information surrounding an event that influences how participants and observers interpret it. Albert Mehrabian's communication model, while with its criticisms, notes that when communicating feelings and attitudes, words make up only a small part of the

communication, with tone of voice and body language making up most of how we communicate. This emphasis on tone of voice and body language is especially relevant for high-context cultures (HCC).

3. Edward T. Hall, *Beyond Culture* (Anchor Books, 1977), 91.
4. Edward T. Hall, *The Silent Language* (Bloomsbury Publishing,1980), 218.
5. Alfons Trompenaars, *Riding the Waves of Culture: Understanding Cultural Diversity in Business* (Economist Books, 1993), 50.
6. The respondents were all managers in multinational and international organizations so may not be representative of the general population.
7. Erin Meyer, *The Culture Map: Breaking Through the Invisible Boundaries of Global Business* (Public Affairs, 2015), 19, 22.
8. Donald A. DePalma and Paul D. O'Mara, *Can't Read, Won't Buy – B2C: Analyzing Consumer Language Preferences and Behaviors in 29 Countries* (CSA Research, 2020), https://insights.csa-research.com/reportaction/305013126/Marketing
9. See Chapter 8 for more on AI and voice interactions.
10. Eric Bellman, "Amazon, to Win in Booming Rural India, Reinvents Itself," *The Wall Street Journal*, December 21, 2018, https://www.wsj.com/articles/amazon-to-win-in-booming-rural-india-reinvents-itself-11546196176
11. Giorgia Lombardo, "A Guide to Cross-Cultural Design — By Senongo Akpem," *DeMagSign*, June 23, 2021. https://medium.com/demagsign/a-guide-to-cross-cultural-design-by-senongo-apkem-368c90de1b76
12. Andrew Ifeanyi Isiguzo, "African Culture and Symbolism: A Rediscovery of the Seam of a Fragmented Identity," in *Symbols, Cultures, and Identities in a Time of Global Interchange*, ed. Paata Chkheidze, Hoang Thi Tho, and Yaroslav Pasko (The Council for Research in Values and Philosophy, 2015), 177. https://www.crvp.org/publications/Series-VII/VII-27.pdf
13. John Cassim, "Zimbabwean Bird Lover Demystifies Myths around Owls," Anadolu Ajansı. Last updated September 1, 2021. https://www.aa.com.tr/en/africa/zimbabwean-bird-lover-demystifies-myths-around-owls/2351204
14. See Chapter 4: Building for Limited Digital Connectivity.
15. The range is from less than 1% to 27%. World Bank. "Owns a Credit Card (% age 15+)." Gender Data Portal. Accessed August 11, 2025. http://genderdata.worldbank.org/en/indicator/fin7-t-a
16. Nyay Bhushan and Georg Szalai, "Netflix Expected to Bring Mobile-Only Plan to More Markets," *The Hollywood Reporter*, January 8, 2020. https://www.hollywoodreporter.com/business/business-news/netflix-expected-bring-mobile-plan-more-markets-1259093/
17. More on this in Chapter 3 and Chapter 7.
18. Covered in greater detail in Chapter 7: The Unique Challenges and Solutions for User Research in Africa.
19. See Chapter 4: Building for Limited Connectivity for more on this.

Bibliography

Bellman, Eric. "Amazon, to Win in Booming Rural India, Reinvents Itself." *The Wall Street Journal*, December 21, 2018. https://www.wsj.com/articles/amazon-to-win-in-booming-rural-india-reinvents-itself-11546196176

Bhushan, Nyay, and Georg Szalai. "Netflix Expected to Bring Mobile-Only Plan to More Markets." *The Hollywood Reporter*, January 8, 2020. https://www.hollywoodreporter.com/business/business-news/netflix-expected-bring-mobile-plan-more-markets-1259093/

Cassim, John. "Zimbabwean Bird Lover Demystifies Myths around Owls." Anadolu Ajansı. Last updated September 1, 2021. https://www.aa.com.tr/en/africa/zimbabwean-bird-lover-demystifies-myths-around-owls/2351204

DePalma, Donald A., and Paul D. O'Mara. *Can't Read, Won't Buy – B2C: Analyzing Consumer Language Preferences and Behaviors in 29 Countries*. CSA Research, 2020. https://insights.csa-research.com/reportaction/305013126/Marketing

Earley, P. C., and Soon Ang. *Cultural Intelligence: Individual Interactions Across Cultures*. Stanford University Press, 2003.

Hall, Edward T. *Beyond Culture*. Anchor Books, 1977.

Hall, Edward T. *The Silent Language*. Bloomsbury Publishing, 1980.

Hofstede, Geert. "The 6-D Model of National Culture." Accessed August 20, 2025. https://geerthofstede.com/culture-geert-hofstede-gert-jan-hofstede/6d-model-of-national-culture/.

Isiguzo, Andrew Ifeanyi. "African Culture and Symbolism: A Rediscovery of the Seam of a Fragmented Identity." In *Symbols, Cultures, and Identities in a Time of Global Interchange*, edited by Paata Chkheidze, Hoang Thi Tho, and Yaroslav Pasko. The Council for Research in Values and Philosophy, 2015. https://www.crvp.org/publications/Series-VII/VII-27.pdf

Lombardo, Giorgia. "A Guide To Cross-Cultural Design — By Senongo Akpem." *DeMagSign*. June 23, 2021. https://medium.com/demagsign/a-guide-to-cross-cultural-design-by-senongo-apkem-368c90de1b76

Meyer, Erin. *The Culture Map: Breaking Through the Invisible Boundaries of Global Business*. PublicAffairs, 2014.

Pecly, Pedro Henrique D., and Priscilla Cristina C. Ribeiro. "Entrepreneurship in the BRICS and Cultural Dimensions." *Brazilian Journal of Operations & Production Management* 17, no. 2 (2020): 1–18. https://doi.org/10.14488/BJOPM.2020.013

The Culture Factor Group. "Country Comparison Tool." Accessed August 28, 2025. https://www.theculturefactor.com/country-comparison-tool

Trompenaars, Alfons. *Riding the Waves of Culture: Understanding Cultural Diversity in Business*. Economist Books, 1993.

World Bank. "Owns a Credit Card (% age 15+)." Gender Data Portal. Accessed August 11, 2025. http://genderdata.worldbank.org/en/indicator/fin7-t-a

Chapter 6

Building Trust in Digital Products

In previous chapters, we looked at sociocultural and technological needs unique to African markets. Product and design decisions that consider these realities facilitate trust, which is necessary for digital adoption.

Misalignment with user perceptions, expectations, norms, and behaviors erodes trust and reduces willingness to use such services. For example, using imagery deemed offensive or failing to provide relevant payment options can cause frustration, a negative perception of the product and the brand, and, ultimately, churn.

Unfamiliarity with a product's features can also impact trust. Users anxious about making mistakes or "breaking something" may hesitate to explore features or use products at all. Clear explanations can help demystify these complexities, creating understanding which builds confidence and reduces apprehension.

Beyond design and usability considerations, ethical and regulatory frameworks also shape trust in digital products.

DOI: 10.1201/9781003648192-8

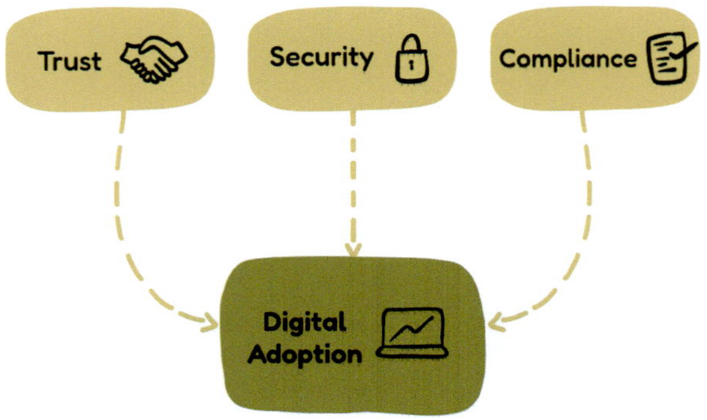

FIGURE 6.1 Achieving digital adoption.

Data Privacy and Security

Privacy and security are often treated as secondary concerns—something to worry about after the product is in users' hands. However, this is a mistake. This approach not only risks reputational damage by breaking users' trust but also exposes companies to financial and regulatory consequences.

Building trust through data privacy and security requires attention to three interconnected elements: complying with regulatory requirements, securing consent and ensuring transparency, and designing robust security measures from the outset.

Regulatory Compliance

Many African countries have enacted data protection regulations similar to the EU's General Data Protection Regulation (GDPR) and to several sector- and regional-level privacy laws in the US, such as the California Consumer Privacy Act (CCPA) and Health Insurance Portability and Accountability Act (HIPAA).

These data protection requirements are often industry-specific, and even within industries, obligations may vary by business model. As Boluwatife Ekundayo notes, "a payment provider requires a different license than a microfinance bank, and a health tech company that also provides medication needs a different license from one that just connects doctors to patients."

Compliance with these regulations and standards is not optional. Even when individuals are not aware of data practices, which is common among audiences with lower digital literacy, the high legal standards must still be met.

While this section focuses on regulations in the Big Four (Egypt, Kenya, Nigeria, South Africa), data protection laws and regulations also exist in Morocco, Mauritius, Senegal, Uganda, Ghana, and other African countries.

Egypt's Personal Data Protection Law No. 151 of 2020 (PDPL)

The PDPL applies to any processing of personal data belonging to individual resident in Egypt, regardless of the location of the data controller or processors. The law is grounded in principles of lawfulness, fairness, and transparency. Data collection must be for specific, legitimate, and declared purposes, and all data must be processed and stored securely.

This regulation grants individuals the rights to access, rectify, erase, and restrict the processing of their personal data, and includes specific provisions for sensitive data, such as health and financial information. It requires explicit consent, especially for processing sensitive data and for direct marketing communications. Data transfer is generally prohibited except in the following cases: the destination country has data protection laws equal to or greater than Egypt's, prior authorization has been obtained from the Egyptian Data Protection Center (EDPC), or the data subject gives explicit consent.

Kenya's Data Protection Act (DPA)

Introduced in 2019, this legislation is built on the principles of lawfulness, fairness, and transparency; purpose limitation; data minimization; accuracy; storage limitation; and integrity and confidentiality.

The Act grants individuals significant rights over their personal data, including the right to be informed, the right to access, the right to object to processing, the right to rectification or erasure, and the right to data portability. Compliance obligations rest with data controllers (entities determining the purpose and means of data processing) and data processors (entities processing data on behalf of controllers). Transferring personal data outside Kenya is restricted, except when the destination country has adequate data protection standards or when other appropriate safeguards are in place.

Nigeria Data Protection Act (NDPA)

The NDPR became law in 2023, replacing the earlier Data Protection Regulation (NDPR) of 2019 as Nigeria's primary legal framework for data privacy. The Act requires lawful, fair, and transparent processing of personal data. Data collection is limited to specific, legitimate, and necessary purposes, and collected data must be accurate and securely stored.

People have the right to be informed about their data, access their data, rectify and erase their data, and object to the processing of their data. Data processors must obtain clear, explicit, and unambiguous consent from data subjects—silence or inactivity does not imply consent. Personal data may be transferred when the country has adequate data protection laws, when the data subject gives consent, or when necessary for the performance of a contract.

EXPERT INSIGHT: BOLUWATIFE EKUNDAYO ON COMPLIANCE REQUIREMENTS

In Nigeria, if you have an AI recommender or chatbot, you have to register with NITDA, no matter what type of service you provide. Businesses involved in data processing and digital operations must register a .ng domain name. There is also tax compliance to consider. Africa is not a country, so you have to do research country by country. For example, Nigeria's Privacy Law does not classify financial data as sensitive data; however, Ghana and South Africa do. It is best to apply the strictest regulations because it keeps you compliant across the board.

South Africa's Protection of Personal Information Act (POPIA)

POPIA came into full effect in 2021 and has eight conditions for lawful processing of personal data, including accountability, processing limitation, purpose specification, information quality, openness, security safeguards, and data subject participation. This regulation applies to the data rights of both "natural persons" (people) and "juristic persons" (such as companies).

Individuals have the right to be notified of data collection, to access their data, to request correction or deletion, and to object to processing. Direct marketing via unsolicited e-communications is prohibited/restricted. Transferring personal data outside South Africa is also prohibited except where the data protection standards in the foreign country are similar to South Africa's, the data subject consents, or other specific conditions apply, such as contractual necessity.

Regional Frameworks

Pan-African data protection initiatives also exist alongside national laws. Two key frameworks are the African Union Convention on Cybersecurity and Personal Data Protection (The Malabo Convention) and the ECOWAS Supplementary Act on Data Protection.

The Malabo Convention

Adopted in 2014, the Malabo Convention aims to create a harmonized legal framework for the continent and is grounded in the principles of lawfulness, fairness, and transparency. The Convention grants data subjects several rights: the right to be informed when personal data is processed; the right to access their data; the right to object to processing personal data on legitimate grounds (especially for direct marketing); the right to request the correction or deletion of personal data if it is inaccurate, incomplete, or unlawfully processed; and the right to file complaints with their country's data protection authority if they feel their rights have been violated.

ECOWAS Supplementary Act

The ECOWAS Supplementary Act A/SA.1/01/10 on Personal Data Protection was adopted in 2010 to provide a legal framework for the 15 member states of the Economic Community of West African States (ECOWAS). The Act establishes baseline standards to harmonize sub-regional data protection laws and facilitate cross-border data transfers. Each member state is expected to implement its own legal framework for privacy protection covering the collection, processing, storage, and use of personal data, based on these baseline standards.

Most ECOWAS member states, including Nigeria, share core principles and requirements. However, it is important to always research the specific country regulations for any differences.

Why Does Regulatory Compliance Matter?

Data breaches—where customer personal data is lost, stolen, or otherwise misused—create not only regulatory liability for the company but also personal risks to individuals. Consequently, users are increasingly factoring data protection practices into their choice of service providers.

Strong data protection fosters customer trust and drives adoption. When users are confident their data is protected and their privacy is not compromised, they are more likely to trust the service provider and use the solution.

It is a myth that only people in high-income countries care about personal data protection. According to the UNCDF, customers in lower-income countries also consider it important and are even willing to pay for greater protection of their personal data or refuse discounts offered in exchange for sharing personal data.[1] Digital services that provide financial, communication, health, or other such services also need to consider the gendered lens of data sharing. UNCDF[2] indicates that women may be less willing to use some digital services due to concerns about data security.

Greater confidence in the safety of their data increases users' likelihood of using the service. Conversely, data misuse and breaches erode customer trust, leading to churn and abandonment. This shows that demonstrating to users that you have robust security measures can be a key selling point.

Consent and Transparency

A common theme across data protection regulations is the requirement for consent and transparency in data collection.

Consent is not given by default. This means people should consciously opt in to sharing specific data necessary for the digital service, rather than being forced to opt out of broad data collection. In some countries, this is a regulatory requirement, not just best practice. South Africa's POPIA, for example, requires opt-in for marketing communications, making unsolicited emails or SMS messages without users' explicit permission a violation of the act.

When seeking consent, request only the data necessary for the user to access the service. There must be a clear, legitimate purpose for every piece of data collected. It should also be easy for users to give, withdraw, or refuse consent; thus, dark patterns (such as coercing users into giving consent or not presenting the option to refuse/withdraw consent) should be avoided.

Users can give consent through granular toggles that allow them to choose which specific data they want to share and with whom. However, be careful not to overwhelm them with too much information. When seeking consent to collect personal data, clearly explain the benefits to users. When people understand why data is needed, how sharing benefits them, and how their information is protected, they are more likely to consent.

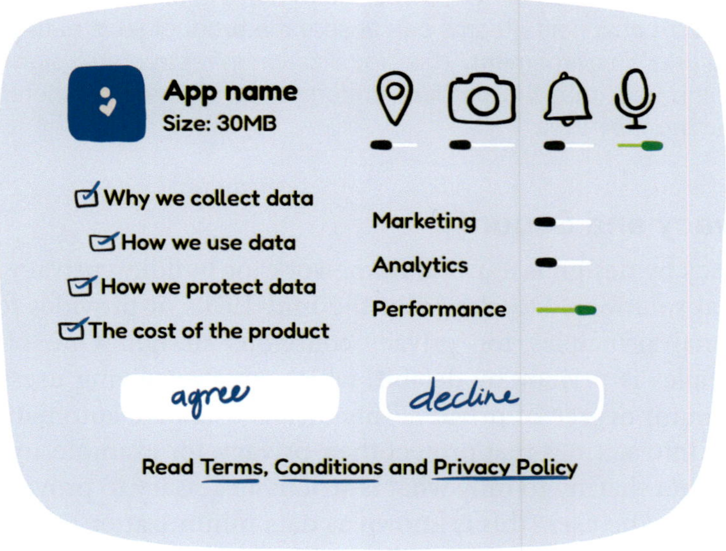

FIGURE 6.2 Transparency in requesting consent.

Users must understand how data is collected, what data is collected, and how the data will be used. However, explanations should be in simple, easy-to-understand terms, not in technical jargon, and use visual aids where helpful.

Privacy policies should be as simple and clear as possible. Research suggests that visual cues, such as illustrations, icons, and comic designs, can help users understand and remember privacy policies, deepen their understanding of privacy, and help them understand what they are consenting to.[3] Other helpful techniques include dividing policies into shorter sections instead of one long block of text and presenting information in ways that don't interrupt users' primary goal.

EXPERT INSIGHT: PRINCESS AKARI ON CONTEXTUAL RELEVANCE

Trust is eroded when a product feels redundant or "foreign" to local needs. Scaling products across countries doesn't work if you only attempt to replicate what has worked somewhere else without adapting to local context. Princess Akari discovered that by adding IP detection to surface only the relevant features per country—such as currency-specific formatting or local customer support channels—teams can ensure the product feels native to the user's environment. This "localization infrastructure" signals to the user that the product understands their reality, thereby earning their trust.

Privacy and Security

Privacy by design is a useful framework for building privacy into digital solutions. Developed in the mid-1990s, it provides foundational principles for privacy-conscious design. One of the principles is privacy by default, which involves giving users the maximum degree of privacy. This means users are automatically opted into settings that protect their privacy, for example, by limiting data sharing to only what is strictly necessary to provide the service to the user (this is known as data minimization).

Privacy by default is especially important in contexts where people share devices, as it reduces risk exposure for users and providers.

In some cases, collecting personal identifiable information (PII) is necessary—financial services providers, for example, must comply with Know Your Customer regulations. Such information should be used only for the purpose users consented to, and that purpose must be clearly communicated.

The same applies to healthcare data. Across the regulations detailed earlier, explicit consent is required to collect or share personal health data except in cases of emergency or within the context of healthcare provider-patient confidentiality. Strict access controls must also govern who can access customer data. For example, using unique identifiers instead of names or email addresses in analytics tools. Health data is sensitive from both regulatory and user perspectives.

EXPERT INSIGHT: BOLUWATIFE EKUNDAYO ON PRIVACY FRAMEWORKS

Companies should implement privacy program frameworks that evolve as the company grows. The best time to start thinking about your privacy program framework is when you start conceiving your idea. You need to create a data map, identify data collection channels, and envisage how the data will be processed. Track how you've used, collected, and stored data over time.

In my view, financial data is sensitive data. My risk index is the severity of harm that could occur if financial data is breached and the likelihood of that harm. It's an easy target for hackers. Under open finance and open banking, there's data interoperability among different providers. APIs enable much of the financial services provision, and companies can call information from my bank. The security risk is that if a provider using my data is compromised and they make a call to my account while hackers have access, the hackers can withdraw my funds. If someone can see what I spend money on, they can profile me. They can see that every Thursday I go to a certain place. If you are collecting card details, then the vulnerability index is high.

Beyond regulatory requirements, users are increasingly cautious about applications that request too many permissions and may choose not to use the product when they don't understand why certain permissions are required. Pew Research reports that 57% of American app users have either not installed an app or uninstalled an app they already downloaded because of concerns about sharing their personal data.[4] A global study also found that when permission requests include an explanation, they are less likely to be declined.[5] Other research shows that people are more likely to install an app if they perceive it as privacy-friendly, expect to benefit from it, and see transparency in data collection.[6] Conversely, they have negative attitudes towards apps that request more permissions than needed, which makes them less likely to download such apps.[7]

Finally, any products that are aimed at, used by, or involving minors must also have parental consent and additional safeguards to protect the privacy of children. In 2023, the Office of the Data Protection Commissioner in Kenya fined a school for posting photos of children without parental consent[8].

Beyond Regulations

Privacy and security, transparency and consent, confidentiality and control—these are basic requirements of many data protection regulations that can serve as a baseline for building trust with users. However, trust is multilayered. It starts with systems and products that work as advertised and as promised. That is the foundational layer. On top of that foundation sit human connection, perceived security, and data use transparency. Users want to feel safe, secure, and at ease. Whether it's about putting their money somewhere, awaiting a delivery, or booking a cab. This means that designers and product builders must incorporate features that promote these feelings. Digital service providers must understand users' fears, concerns, and hesitations, then address them by meeting people where they are and designing trust-building mechanisms that reduce user anxiety. The Silicon Valley ethos of moving fast and breaking things does not work in this context. As Odun Eweniyi, one of the cofounders of Piggyvest put it: "move thoughtfully and don't break trust."

Human-Centered Validation

Trust is a key factor for many people, even those who are not new to the internet. Building this trust goes beyond promises of privacy and transparency on app screens—it is built by doing things that don't scale, especially in low-trust societies and places with high loss aversion and uncertainty avoidance. In short, products that use community-based agent networks to build trust drive adoption faster.

Part of Amazon's growth in India is attributed to building infrastructure that supported people new to online shopping, who often didn't have credit cards or delivery addresses they could fill online.[9] Amazon partnered with existing neighborhood stores, integrating them into its distribution network as package depots or learning centers. The store owners were already part of the community, so the residents trusted them.

This is similar to the agent network model that companies like OPay, Moniepoint, and Safaricom have used across Africa to reach new users with limited access to smartphones and the internet. It also reflects how many people in remote areas learn to use the internet from trusted people in their community who help them figure out difficult steps.

You don't have to create an agent network to build trust. Companies like Piggyvest built trust by making the founders visible to the community of users they were trying to build. The product also spread through word of mouth as early adopters who had experienced it spoke to those in their network.

EXPERT INSIGHT: YANN LE BEUX ON DIGITIZING BEFORE THE LAST MILE

Success often comes from digitizing just before the "last mile" rather than the end user. Instead of trying to onboard a rural farmer who has connectivity problems onto a complex app, builders should provide digital tools to the "Extension Agent" or "Community Health Worker."

You don't necessarily pick the elderly community leaders; you pick younger, tech-savvy people to act as powered agents. This model is often criticized by VCs for not being "scalable" in the traditional sense, yet it is more effective at achieving growth and delivering value than apps that attempt to onboard rural users directly.

> In every agriculture project we at YUX Design have seen, trying to onboard farmers on a digital app often fails to some extent. But as soon as the model was to create a network of agents that had a mobile phone and an app, it worked better.

Similarly, concepts of digitizing before the last mile use tech-savvy "Extension Agents" or "Community Health Workers" to bridge the digital literacy gap, leveraging a trusted human intermediary instead of forcing a complex app onto a new digital user.

Increase Perceived Safety and Security

Trust is essential for digital adoption and is built over time. Reliability and continued successful encounters—in functionality and in experiences—increase the user's willingness to do more activities or use more advanced features as they go along. For example, users may become more comfortable with the idea of adding their card to a service after multiple successful transactions—from purchase or payment to receiving their goods or services—over time.

EXPERT INSIGHT: TOYIB MUSTAPHA ON TRUST IN ONLINE PAYMENTS

Users often prioritize perceived security over convenience. For instance, while product teams may push for automated, saved card payments for efficiency, many users remain skeptical. Users are often reluctant to complete the process, concerned about a third-party service "holding on to their card." This lack of trust is often overcome by offering alternatives, such as manual bank transfers. In this context, users may prefer the control and visibility of transferring funds for each transaction rather than risking an automated payment.

Trust may also be tied to a user's sense of personal safety. Ride-hailing apps like Bolt, Uber, and InDrive, which have implemented safety and security features like sending a push notification to check in on the welfare of riders when the vehicle has been in the same position for too long, are a good example to follow. It could also look like SOS buttons, in-app driver verification, and trip-sharing functionalities. These features build user trust in an environment where personal security can be a concern.

Another thing that builds trust is when users feel mentally at ease. Food delivery apps like ChowDeck provide real-time tracking to users so they know when to expect their food, removing anxiety about where their order is and when to expect it.

Even features such as biometric or other forms of authentication, like pins and passwords, can reduce users' anxiety about the privacy of their data.

Data Use Transparency

Transparency extends to costs and fees. You may immediately think of subscription fees or in-app purchases, but as discussed in previous chapters, this could also include data costs. Because of the high cost of mobile data, users are conscious about how much data apps and features consume. Clearly communicating this will help build trust. Yann Le Beux points out that infrastructure challenges often force users to "default to turning off their cameras" during Zoom calls or avoiding high-def settings. If your app doesn't inform the user of the "weight" of a task (in MBs), you are likely creating a "robbed" feeling that erodes trust.

Provide accurate information about what users can expect, do, or accomplish with your service. Setting realistic expectations in this way provides a better experience than overpromising features or benefits. Similarly, avoid presenting features unavailable in a user's location. This could lead to frustration and a sense of exclusion, which undermines trust in the product.

MUTUGI KIREMA: A CASE STUDY ON FANCY ANIMATIONS AND HIGH DATA COSTS

In one of my projects back in 2023 for a top-tier East African bank, we initially designed an animated banner for the home screen. It rotated through credit offers and savings prompts, giving the app a sleek, modern feel. However, during user research across Kenya, Uganda, Rwanda, and Tanzania, we quickly learned that what looked beautiful in the design studio wasn't practical in the real world. Many users were on entry-level smartphones with limited memory and expensive mobile data plans. Some even said, "Every second the app loads, I'm counting my bundles."

We ran field tests to measure data usage and app performance under typical network conditions. The banner alone consumed up to 30% of the total data per session. That insight forced us to strip it back completely. We replaced it with a static image and introduced a "Low Data Mode" option that allowed users to turn off rich media altogether. The result? A 33% drop in data consumption and a significant reduction in app abandonment during loading.

The real balance came from understanding that visual trust isn't built solely through animations or high-res imagery. It's also built through clarity, speed, and respect for the user's resources. We retained strong brand colors, simplified icons, and crisp typography to keep the experience elegant yet lightweight. Ultimately, prioritizing efficiency over flair made the product more inclusive.

Privacy-Centered Features

Privacy in this context goes beyond data collection and management to include the realities of shared devices, sideloaded applications, shared email accounts, and other factors that expose users to privacy and security risks. Not everyone owns a device or has an email address, so requiring email addresses for product access could create barriers and may lead users to share accounts or use friends' credentials, introducing privacy and security risks. As discussed in other chapters, phone numbers are more widespread and can serve as an alternative to email for account creation or access.

Interface design also requires privacy considerations. In voice-enabled applications, for example, users may prefer visual or text responses to audio responses, especially in public spaces. Voice-first or voice-only features may not be ideal in these circumstances. At the same time, because of limited literacy, offering voice interactions could reduce barriers to access. As we've noted before, the key is to give users the option and allow them to decide how they want to interact with a product or service.

Maintaining Privacy with Shared Devices

As discussed in previous chapters, there are scenarios in which users share devices, email accounts, or even phone numbers. This could be a shared smartphone at home, a shared tablet at work, or a shared computer at a local cafe, library, or other public-facing institution or environment. Device sharing raises privacy concerns, including unauthorized access to personal information (e.g., health data) and accidental (or deliberate) alterations or deletions of sensitive information.

These risks extend beyond personal use to corporate settings. For example, hotel or restaurant frontline employees may need to access personal online training resources on a shared device, potentially exposing their personal information to coworkers or managers.

EXPERT INSIGHT: YANN LE BEUX ON PRIVACY

Shared device usage often creates tension between support and privacy. Children may help their parents navigate a mobile money app, but builders must design safeguards to prevent the "helper" from accessing funds without permission. Institutional gatekeepers can also sometimes act as "roadblocks" to digitization. Digitization brings transparency, which can eliminate "negotiation fees" that happen along the value chain. Building trust requires showing these leaders what they gain from a digital system, rather than just what they lose in control.

A simple solution is to enable multiple user profiles, each password-protected. Each user can have their own preferences, privacy settings, and so on. The considerations for designing the switching mechanisms are the same as those for designing any digital tool. Profile switching should be easily discoverable, and the current profile should always be clear to users. Privacy should be the default.

Multi-profile support is not entirely new. Many software (and even hardware) providers already offer this. For example, Netflix allows viewing profiles, and users can set a PIN so that others

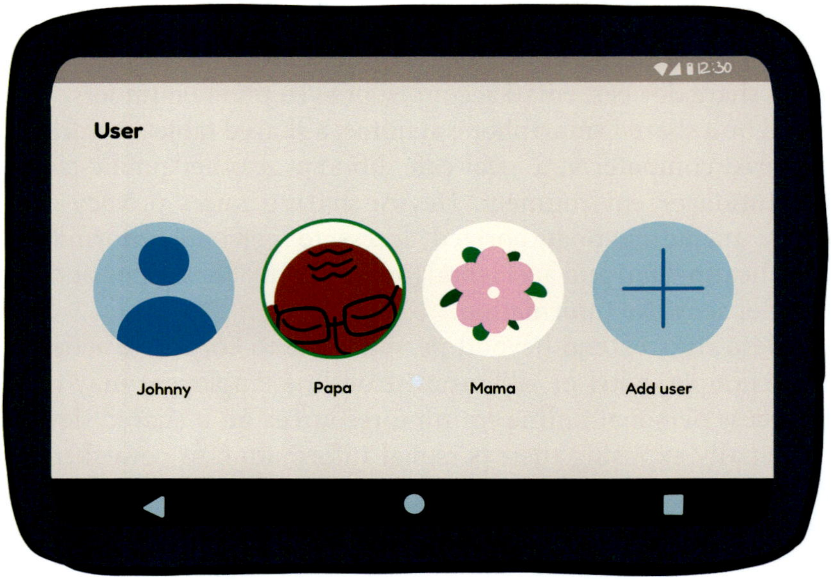

FIGURE 6.3 Multi-profile support.

can't access their own profile. Android devices often have multiple user profiles. However, as Jerahmeel Madumere, an Information Security Researcher, points out, the feature is often hard to discover, especially for users with lower digital literacy levels, as it is buried in complex device settings. More can also be done to improve discovery at the device level and introduce it at the app or service level.

Ethical Considerations in Research

As established, conducting thorough research to understand users and their surrounding environmental and sociocultural context helps move beyond stereotypes. However, effective research requires careful attention to ethics. This is important because localized research is a way to earn a first layer of trust. Before trust is established via features such as regulatory compliance and security, it is first built in the creation process. Insights gained through ethical research practices enable founders and product teams to make decisions that power trust. If the research process—a point of human connection between participant and product—is not conducted with respect and care, it can negatively impact the level of trust users have in the product or service.

Basics of Ethical Research

The principles of privacy and data protection discussed earlier in this chapter apply equally to research. When conducting research, participants must fully understand what the research entails. Be transparent about the purpose, methods, and intended use of research data. Participants should know they can withdraw consent at any time without penalty. Sessions should never be recorded without opt-in consent, and the request should give participants the option to say no to being recorded.

Minimization of data collection, anonymization by removing personally identifiable information, and secure storage of research data must also be implemented. Data retention timelines must be adhered to. For example, if you tell participants their data will be deleted after 30 days, it must be deleted—not anonymized and retained. Anonymizing the data is a violation.

EXPERT INSIGHT: BOLUWATIFE EKUNDAYO ON PROCESSING RESEARCH DATA

Customer data might be collected on the basis of legitimate interest. To improve services, companies may need to get feedback from existing clientele. However, the moment they withdraw consent, you must stop processing their data. So all the information you've collected must be deleted. Even if they withdraw consent before you analyze the data.

Furthermore, there must always be a record of when and how consent was given and/or withdrawn.

Researcher Effect

The researcher's identity—for example, race, gender, and perceived social class—and their position relative to the participant could affect participant behavior and disclosure, thereby impacting the validity of the data. Participants may filter their answers to conform to what they believe the researcher wants to hear. For example, in a culture where women don't discuss certain issues in public, a male researcher's presence could limit their willingness or ability to answer questions fully or truthfully. These differences could also impact trust and openness, especially around sensitive issues.

In digital product design, researchers often represent technology companies that participants may perceive as powerful. This creates a power imbalance, especially with vulnerable or marginalized groups, where participants may feel like they can't criticize or make "negative" comments about the product experience.

One way to address this is to practice reflexivity: researchers should recognize their own position and how it might impact the data collected and how it is analyzed. Potential biases should be documented in the data analysis. Working with researchers or product team members local to the context, culture, or environment can also help address these dynamics.

Incentives

Offering incentives is standard practice in research and is recommended when working with vulnerable populations. However, think critically about the type, value, and structure of the incentives to avoid undue influence or perceived coercion.

A large incentive can weaken a participant's ability to voluntarily consent to research participation, especially when there are risks. Excessive incentives could exploit the economically vulnerable and influence how participants respond—they may think their responses affect whether or how much they will be compensated.

To prevent this, incentives should be commensurate with the time and effort participants have to give. Inform participants about the incentive value and conditions before obtaining consent. It is also preferable to avoid performance-based incentives; compensation should be based solely on participation.

Key Takeaways

1. Trust is the primary currency for adoption, and physical presence through human-centered validation is essential in risk-averse, low-trust environments
2. Builders must comply with regulations as a competitive advantage, signaling legitimacy to users
3. Transparency, privacy, and informed consent are essential to build user trust and must be integrated from the start and not be an afterthought
4. Because shared devices are common, products must mitigate privacy risks by providing features that prevent unauthorized access to sensitive data

What Next?

Reflection Prompts

Within your team, reflect on the following questions, highlighting where your team is doing well and brainstorming ways of fixing whatever problems you uncover.

Reflection prompt	Why it matters
If our digital product disappeared tomorrow, how can users reach out to us?	*Users need a human fallback. If your app fails, is there a physical agent, phone number, or other person/resource they can turn to*
Is our default setting protecting the user?	*Users should not have to dig through settings to be safe*
If multiple people are using the device that our user is accessing our app through, is their data protected?	*Users who share devices are at risk of their sensitive data being exposed to others*
Do we have a privacy program framework with a comprehensive data map of every piece of customer data collected and metadata about it?	*Proactively thinking about a privacy program even from idea conception helps prevent regulatory issues in the long term*
For every piece of data we request, have we established and communicated a clear, legitimate purpose, and are we requesting only the data that is strictly necessary for the user to access the service?	*Regulations, customer preferences, and privacy best practices mean that it is important to only request for what is needed*

(*Continued*)

Reflection prompt	Why it matters
Do we actively demonstrate to users that we have robust security measures?	*This can be a key selling point for users who are concerned about their data security and privacy*
When seeking consent, do we clearly explain the benefits to users?	*Clear and easy to understand explanations help users make informed decisions*

Ethical Research Protocol Worksheet

Use this to assess your research plan for trust beyond the basics.

Ethical check	Action item
Incentives	Is the incentive fair but not coercive? (Avoid excessive amounts that force participation from vulnerable groups.)
Withdrawal	Have we made it clear that users can stop at any time without losing their incentive?
Data retention	Do we have a specific date to delete (not just anonymize) the recordings?
Transparency	Have we explained who we are and how this data helps them/the community?

Notes

1. United Nations Capital Development Fund, *The Role of Data Protection in the Digital Economy* (UNCDF, 2021), 3. https://policyaccelerator.uncdf.org/all/brief-data-protection-digital-economy
2. United Nations Capital Development Fund. *The Role of Data Protection in the Digital Economy* (UNCDF, 2021), 4. https://policyaccelerator.uncdf.org/all/brief-data-protection-digital-economy
3. Agnieszka Kitkowska et al., "Designing for Privacy: Exploring the Influence of Affect and Individual Characteristics on Users' Interactions with Privacy Policies," *Computers & Security* 134 (September 2023): 103468. https://doi.org/10.1016/j.cose.2023.103468
4. Jan L. Boyles, Aaron Smith, and Mary Madden, "Privacy and Data Management on Mobile Devices," *Pew Research Center*, September 5, 2012. https://www.pewresearch.org/internet/2012/09/05/privacy-and-data-management-on-mobile-devices/

5. Weicheng Cao et al., "A Large Scale Study of User Behavior, Expectations and Engagement with Android Permissions," *30th USENIX Security Symposium (USENIX Security 21)* (2021): 803–20. https://www.usenix.org/conference/usenixsecurity21/presentation/cao-weicheng
6. Florian Bemmann and Sven Mayer, "The Impact of Data Privacy on Users' Smartphone App Adoption Decisions," *Proceedings of the ACM on Human-Computer Interaction* 8, no. MHCI, Article 278 (September 2024): 1–23. https://doi.org/10.1145/3676525
7. Hsiangchu Lai, Jack S. Hsu, and Min-Xun Wu, "The Impacts of Requested Permission on Mobile App Adoption: The Insights Based on an Experiment in Taiwan," *Proceedings of the 51st Hawaii International Conference on System Sciences* (2018). https://doi.org/10.24251/hicss.2018.490
8. Post from the Official LinkedIn page of the Office of the Data Protection Commissioner. https://www.linkedin.com/feed/update/urn:li:activity:7112391108491395074/
9. Eric Bellman, "Amazon, to Win in Booming Rural India, Reinvents Itself," *Wall Street Journal*, December 21, 2018. https://www.wsj.com/articles/amazon-to-win-in-booming-rural-india-reinvents-itself-11546196176

References

Bellman, Eric. "Amazon, to Win in Booming Rural India, Reinvents Itself." *Wall Street Journal*, December 21, 2018. https://www.wsj.com/articles/amazon-to-win-in-booming-rural-india-reinvents-itself-11546196176

Bemmann, Florian, and Sven Mayer. "The Impact of Data Privacy on Users' Smartphone App Adoption Decisions." *Proceedings of the ACM on Human-Computer Interaction* 8, no. MHCI, Article 278 (September 2024): 1–23. https://doi.org/10.1145/3676525

Boyles, Jan L., Aaron Smith, and Mary Madden. "Privacy and Data Management on Mobile Devices." *Pew Research Center*, September 5, 2012. https://www.pewresearch.org/internet/2012/09/05/privacy-and-data-management-on-mobile-devices/

Cao, Weicheng, Chunqiu Xia, Sai T. Peddinti, David Lie, Nina Taft, and Lisa M. Austin. "A Large Scale Study of User Behavior, Expectations and Engagement with Android Permissions." *30th USENIX Security Symposium (USENIX Security 21)* (2021): 803–20. https://www.usenix.org/conference/usenixsecurity21/presentation/cao-weicheng

Kitkowska, Agnieszka, Yefim Shulman, Leonardo A. Martucci, and Erik Wästlund. "Designing for Privacy: Exploring the Influence of Affect and Individual Characteristics on Users' Interactions with Privacy Policies." *Computers & Security* 134 (September 2023): 103468. https://doi.org/10.1016/j.cose.2023.103468

Lai, Hsiangchu, Jack S. Hsu, and Min-Xun Wu. "The Impacts of Requested Permission on Mobile App Adoption: The Insights Based on an Experiment in Taiwan." *Proceedings of the 51st Hawaii International Conference on System Sciences* (2018). https://doi.org/10.24251/hicss.2018.490

United Nations Capital Development Fund. *The Role of Data Protection in the Digital Economy.* UNCDF, 2021. https://policyaccelerator.uncdf.org/all/brief-data-protection-digital-economy

Part 3

Looking Forward

Chapter 7

The Unique Challenges and Solutions for User Research in Africa

Hillary's foray into user research was born from pure frustration almost a decade ago. After moving back to Nigeria following five years in the US for university, she immediately noticed a difference. In the US, apps had simply worked. Apple Music was a daily companion—tap, play, done. Back home, that ease disappeared.

At least four Nigerian bank cards failed when she tried to pay for a monthly Apple Music subscription. Eventually, she gave up and continued using a US card, but the problems didn't stop there. Downloading music the way she used to was no longer possible. Three songs could drain a 25GB data bundle within hours. Streaming wasn't much better because songs paused every few seconds, even though, like many others working remotely, she believed she had the best internet money could buy at the time.

Meanwhile, Bozoma Saint John was on stage at Apple's WWDC, celebrating the new Apple Music interface, dancing to Afrobeats, and presenting a frictionless experience to the world. That version of Apple Music felt worlds away from the one loading endlessly in Lagos.

DOI: 10.1201/9781003648192-10

FIGURE 7.1 Testing an app.

The frustration sparked curiosity. Hillary started testing other apps and websites, global and local alike, and involved her mum in the process. Where services felt difficult for Hillary to navigate, they were nearly unusable for her mum. Yet many of these same companies publicly claimed they were "doubling down" on Africa.

Weeks of investigation eventually led to a discovery: there was an entire field dedicated to understanding user needs and pain points before building products. That field was user research.

Going beyond Surface-Level Adaptation

Looking back, Apple Music wasn't badly designed—it just wasn't designed with people like us in mind. It assumed fast data, global card access, stable networks, and a particular user journey. Apple Music, like many products entering the sub-Saharan African market, likely relied on localization strategies, such as tailoring content to reflect cultural preferences—for instance, Top 100 Nigeria or Top 100 South Africa charts. But adapting content to local tastes is surface-level and insufficient on its own.

Chui Chui Tan, a researcher and author of several books on international user research, popularized the term "culturalization," which she describes as going far beyond localization and "infusing cultural authenticity and depth into a product or service."[1] Tan introduces three levels of culturalization in her framework: respect establishment, cultural expectation, and experience enhancement.[2]

Whereas localization focuses on surface-level research to meet cultural norms and legal requirements, culturalization goes deeper. It looks at cultural sensitivities, beliefs, values, everyday habits, and lived contexts, including infrastructure realities. When products don't account for these nuances, they simply cannot succeed in African markets.

This distinction is central to the concept of hyper-local customization, which we introduce as a core element of the Building for Sociocultural Context pillar of the African Product Resilience (APR) Framework™. Culturalization provides the vision—the deep, empathetic understanding that Africa is not an emerging version of somewhere else. Hyper-local customization is the practice—the intentional, grassroots effort to adapt every aspect of the product to specific, on-the-ground realities, moving far beyond mere language translation.

If culturalization demands a deep understanding of cultural context, infrastructure realities, and lived experiences, how can researchers achieve that understanding? Not through the methods most are taught. The user research methods we learn in bootcamps or graduate schools—clean-room usability tests and remote surveys sent via email—are built on a set of assumptions: that infrastructure is reliable, that users are highly literate, and that the relationship between the researcher and the participant is transactionally neutral. In African markets, these assumptions collapse.

These assumptions are immediately challenged by the on-the-ground reality of African markets. The continent is not a single market but a fragmented landscape of over 54 countries and thousands of languages, directly contradicting expectations of reliable infrastructure and uniform user literacy. Furthermore, product teams learn quickly that discovery work cannot be separated from the reality of systemic mistrust and deep cultural nuance—factors that undermine the notion of a "transactionally neutral" relationship between researcher and participant.

We've established that standard research methods fail in African contexts because they assume conditions that don't exist. Developing contextually appropriate approaches requires significant adaptation. Before outlining the African-specific research approaches required for deep culturalization, let's examine the fundamental value and true purpose of user research when building for this complex market.

What User Research Actually Does

A common misconception is that user research only serves the product being developed or improved. People think of it narrowly: validate the flow and ship it. While user research helps teams understand the needs for a specific product, its impact is broader.

When you truly understand cultural nuances, user behaviors, economic realities, and trust dynamics, that knowledge ripples through nearly every part of a business:

- **Growth teams** learn which channels people trust, rather than just which ad slots are cheapest
- **Operations teams** map how users move through physical and digital touchpoints
- **Social media** teams identify what cultural moments matter and which references might backfire
- **Customer support** can anticipate where users will get stuck or confused and prepare helpful resources before support tickets flood the helpdesk

User research also informs strategic business decisions that shape everything else:

- Defining your north star metric (this includes daily active users or something more contextual, like successful transactions per week)
- Building payment infrastructure that matches how people access money (mobile money vs. bank transfers vs. agent cash-out)

- Pricing strategy that considers local purchasing power (monthly subscriptions vs. pay-as-you-go vs. micro-transactions)
- Market entry decisions (which cities or countries to launch in, and in what order)
- Partnership strategy (which telcos, agents, or platforms to integrate with)
- Regulatory compliance (how to meet KYC requirements without creating friction)

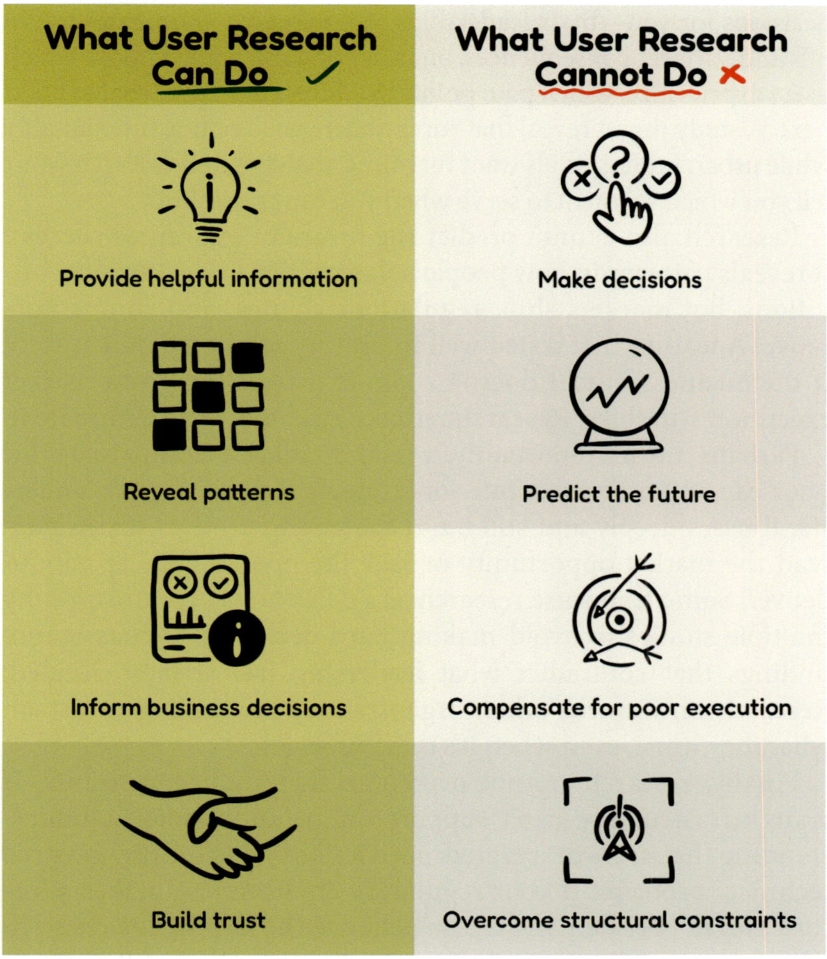

FIGURE 7.2 User research, its scope versus its limits.

Importantly, user research extends beyond primary research activities, such as surveys, user interviews, usability testing, focus groups, and diary studies. It also includes desk or secondary research, such as social listening, keyword research, competitor analysis, internal data like customer support logs (for existing businesses), and academic research, including peer-reviewed articles.

What User Research Cannot Do

For all its value, user research has clear limitations. It cannot make decisions for you—that's leadership's job. Research illuminates trade-offs and surfaces consequences, but someone still has to choose which users to prioritize, which pain points to address first, and what to build next. A study might reveal that rural traders need offline functionality while urban professionals want real-time analytics, but research won't tell you which segment to serve when you can't serve both.

Research also cannot predict the future or guarantee success. It reveals patterns in how people behave today under current conditions, but markets shift, regulations change, and competitors move. A feature that tested well in prototype can still fail at scale if the business model doesn't support it or if the go-to-market execution stumbles. Research reduces risk; it doesn't eliminate it.

Perhaps most importantly, research cannot compensate for poor execution or substitute for strategic vision. You can understand users deeply and still build the wrong product if you misread the market opportunity or lack the operational capacity to deliver. Some teams use research as a delay tactic, commissioning multiple studies to avoid making hard decisions. Others ignore findings that contradict what leadership has already decided. Research only works when organizations are willing to act on what they learn, even when it's uncomfortable.

Finally, research cannot overcome structural constraints. If your infrastructure can't support offline functionality, understanding that users desperately need it doesn't magically solve the technical problem. If your regulatory environment makes a feature illegal, user demand doesn't change the law. While research tells you what matters and why, building solutions still requires resources, time, technical capability, and sometimes luck.

What research *can* offer in African contexts is the ability to build trust. And in environments where trust is limited and hard-won, that impact often matters just as much as the insights themselves.

Contextual Friction: Building Trust before Insight

Most product teams approach research as a straightforward exchange: they ask questions, participants provide answers, and the team uses those responses to build features. In many African markets, however, this dynamic doesn't hold. The hesitation people express rarely stems from the research method itself; it stems from uncertainty about who the researcher represents and how the information will be used. Participants are often assessing whether sharing their experiences exposes them to any form of risk, whether financial, social, or administrative.

This reframes the starting point of research. Before learning about product behaviors, researchers must establish why participants should feel safe enough to speak or share information honestly. A long history of institutional instability and frequent digital scams in the region means trust plays a fundamental role in how people evaluate both products and researchers.

In fintech, this is even more pronounced. As Seyi Oniyitan, a product designer at RiseVest, said, "*The moment you announce you're a fintech, you already have baggage.*" That baggage is visible in both perception and experience. Survey data suggests that, on average, only around four in ten polled Africans express trust in key formal institutions.[3] Africa is also among the regions most exposed to digital fraud worldwide. In 2025, the region scored 3.84 on the Global Fraud Index, significantly higher than the global average of 2.79.[4] Across several African countries, customer experience research suggests that failed transactions and network downtime lead to issues that cause distrust of mobile money and other digital financial services.[5] This backdrop of institutional mistrust and operational friction shapes every interaction participants have with financial products.

In some contexts, researchers may be perceived as authority figures which can lead to distorted responses. Deference to authority and politeness norms can shape data quality, particularly when researchers are visibly foreign, urban, or formally educated. Research shows that self report measures are prone to social desirability bias, which can distort findings.[6]

These same dynamics shape how research itself is interpreted. Participants may question whether a researcher is connected to government authorities, a foreign organization likely to leave abruptly, a competitor, or an NGO collecting data without follow-through. Oluchi Audu, a design researcher at YUX Design, has observed similar patterns in her work in Northern Nigeria, where people were willing to speak but often asked about previous teams that had collected stories and never returned. This echoes the concept of "research fatigue" outlined by Cherop et al., where Kenyan communities repeatedly studied become less willing to engage and less open in their responses.[7] People are also less willing to participate in research when they are concerned about surveillance and the safety of their data, don't see any tangible benefit to themselves, or believe that participation can have negative consequences for them.[8] Research on data collection in Kenya shows that responses may vary both because interviewers differ in how they administer surveys and because respondents react differently to them based on trust, familiarity, or other perceived characteristics.[9]

This helps explain why conventional research methods often perform poorly. Surveys assume confidence in confidentiality, even though many people worry their responses could be seen by authorities or community members. Focus groups assume people can speak candidly in front of peers, yet research in collectivist settings shows that hierarchy, age, and gender shape what is said publicly. Video-based research is also limited: network unreliability in sub-Saharan Africa frequently disrupts digital communication, including video calls. Even when calls do go through, participants often refuse screen recording, particularly during financial tasks, due to concerns about where the footage might be stored or shared. In practice, trust operates across multiple, overlapping dimensions that researchers must navigate simultaneously, such as:

- Perceived Safety of the Research Encounter: This refers to whether the participant feels comfortable sharing information with the particular researcher, in the setting (location) they're in, and in that exact moment. Research in Kenya shows that interviewer identity can materially affect participant responses[10]

- Confidence in Confidentiality: Do participants believe their responses will remain private? In one study in Ghana, respondents cited fears of unauthorized access and data being used against them; this mistrust led many to limit what data they shared with data controllers or to disengage from data-collection exercises altogether.[11] Without credible assurance, responses became guarded or strategically curated
- Understanding the Research Purpose: Can participants grasp why they're being asked certain questions and how the information will be used? When people have concerns about how their data will be used, they are hesitant to participate in research. This has been seen in refugee communities[12] and can easily be the case elsewhere
- Researcher Legitimacy: Can the researcher demonstrate genuine connection to the community or a credible endorsement from trusted figures? Research on M-Pesa adoption showed that trust transferred through networks—when community members or respected neighbors vouched for something, others felt safe engaging.[13] The same dynamic applies to research. A researcher introduced by a trade association leader gains immediate credibility that no company badge can provide. Without this social proof, researchers remain outsiders whose motives are suspect
- Comfort with the Research Method: Do participants feel capable and safe using the research format being offered? Phone interviews might feel safer than video calls, and WhatsApp voice notes might feel safer than live calls. Face-to-face conversations in familiar settings might be the only format where participants fully relax

The implications for research practice are substantial. Researchers cannot simply arrive, recruit participants, and expect valid data. Building trust requires deliberate effort: working with community gatekeepers who can vouch for legitimacy, using participatory approaches that position community members as co-researchers rather than subjects, returning to the same communities multiple times to build familiarity, being transparent about what the research can and cannot achieve, and adapting methods to what feels safe rather than what's most convenient.

Without that foundation, the data is often misleading. People will still participate, but their responses may be cautious, designed to avoid trouble, shaped by what they think researchers want to hear.

Adapting Research Methods for African Contexts

The teams that succeed in African contexts don't try to force traditional user research methods to function under conditions they were not designed for. Approaches such as random sampling, remote usability testing, and transactional researcher–participant relationships assume high institutional trust and stable infrastructure, but these conditions often don't exist. Effective teams build and adapt methods that treat communities as co-creators and context as non-negotiable.

When research relies only on online surveys or digital recruitment panels, it captures the perspectives of a connected urban minority. Understanding the majority—those operating in informal economies, rural areas, or low-connectivity environments—requires different approaches from the outset:

Recruiting through Trust Networks

Traditional random sampling is often ineffective, too costly, or simply impossible to execute when trying to reach users who operate outside formal systems (the unbanked, traders, gig workers). Instead, recruitment hinges on trust networks. This means embracing methods that prioritize legitimacy through community endorsement:

- Snowball Sampling: Participants introduce the researcher to someone they trust, reducing the social risk of participation and helping teams gather the insights they need. Purposive sampling approaches of a homogeneous groups

can yield useful results and help uncover most key themes within 6–12 interviews[14]
- Community Gatekeepers: Participants are recruited through trusted intermediaries such as WhatsApp group admins, faith-based leaders, youth coordinators, market association heads, or local entrepreneurs. These gatekeepers lend credibility and help researchers access groups that would otherwise be unreachable

Participatory Research as Access Pass

One of the strongest adaptations is treating users as active contributors rather than subjects. This approach helps overcome mistrust, power distance, and research fatigue.

> YUX Design's project for UNICEF in Sokoto and Katsina illustrates this clearly. They knew that entering conservative communities as outsiders would lead to polite but shallow responses. So the team recruited and trained thirty young women in each state to help conduct the interviews, run surveys, and shape early ideas. They were treated as part of the research team, rather than as study participants. This immediately changed the dynamics: people opened up more and spoke more honestly about restrictions on phone use and daily routines. Some of the young women kept working on ideas even after the project ended because they felt invested in the work.

This approach produces two critical patterns:

- Lower Power Distance: People disclose more when speaking with peers—"people like me"—and feel safer discussing sensitive issues such as gender norms, mobility, safety, or income
- Sustained Ownership: Solutions are more grounded in reality, and communities remain engaged beyond the research period

These patterns appear across the region. The PEER method (Participatory Ethnographic Evaluation and Research) has been particularly effective in fragile or low-trust environments. In South Sudan, marginalized women were trained to design research tools

FIGURE 7.3 Participatory user research session with community members in a rural village.

and collect and interpret data on maternal health.[15] PEER allowed researchers to gather rich, contextual insights in a fraction of the time traditional ethnography requires.

Other Participatory Methods

Participatory practices extend beyond co-researcher models. Teams across the continent use methods such as:

- Contextual Immersion: Researchers spend time in the community, often two to six weeks in rapid assessments, observing routines, learning norms, and building rapport before conducting formal interviews. While shorter than traditional ethnography, these immersions still provide enough depth to reveal patterns that remote methods cannot
- Community Validation Sessions: Researchers return to share preliminary insights or early prototypes, inviting critique and refinement. These sessions help ensure solutions remain grounded in actual needs and expectations

- Cultural Probes: Giving participants cameras, journals, or simple recording devices to document their own experiences over time. This allows researchers to capture moments and behaviors that would otherwise be missed

Asking Sideways to Find the Truth

A core tenet of discovery research is to avoid leading the user. Yet, in many African contexts, a direct question may be met with polite silence or an answer that the participant believes the researcher wants to hear. To bypass these cultural politeness norms and uncover sensitive truths, people building for or researching the region often have to learn to ask sideways.

YUX Design found great success using indirect, projective methods to address sensitive issues like shared phone use or gender restrictions. They would use printed photo cards showing everyday scenes, or craft short stories with themes like *"Aisha wants to learn sewing on YouTube, but her husband tells her to stay home."* When participants were asked if they related to Aisha's story, many interrupted, saying, "That's me!" or "My friend too!"

Talking about someone else is safer than talking about oneself. This psychological projection creates emotional distance, giving users permission to share their real experiences without fear of judgment. This allows researchers to gather rich qualitative data that a direct question would have shut down.

The Value and Limits of Focus Groups

According to Kome Sideso, Deputy Head of CcHub Design Lab in Rwanda, focus groups function very differently in East Africa compared to West Africa. In Kenya, Uganda, and Tanzania, focus groups often outperform individual interviews because participants are more comfortable discussing experiences collectively. Social norms encourage consensus-building and shared reflection. People validate one another's stories, reveal contradictions, and build on each other's points, which leads to richer insights.

In contrast, in many West African contexts, participants may default to deference and allow dominant voices to take over or hesitate to contradict others publicly. Understanding these cultural dynamics helps researchers choose when a focus group is appropriate and when it may distort findings.

Counting the Offline Majority

Elite bias—data that over-represents the urban, educated, and digitally connected—affects quantitative research just as much as qualitative. To correct this, researchers at YUX rely on enumerator-led surveys: trained local field teams who collect responses in person across informal markets and low-connectivity areas. Enumerators explain questions, support comprehension, and capture insights from people unlikely to complete an online survey.

Research teams across Nigeria, Kenya, and Senegal repeatedly describe this approach as essential for reaching the "offline majority"—populations central to the success or failure of digital products but invisible in digital datasets.

Infrastructure and Regulatory Constraints That Shape Research Methods

Princess Akari, a product manager who has worked on fintech products at Brass and Moniepoint, learned early how fragile traditional research plans can be when they face real infrastructure conditions. Her studies were neatly set up: scheduled Zoom interviews, detailed guides, screen recordings, and well-defined tasks. But when she began running sessions, she spent most of her time hearing, "Can you hear me?" as calls dropped mid-sentence and participants disappeared for minutes at a time or insisted on joining with audio alone to save data. After several sessions fell apart, she stopped trying to "fix" participants and instead rethought her entire approach.

She rebuilt her process around the assumption that infrastructure will fail at some point. Now she plans every study with a "modalities ladder." In-person sessions come first when possible, and video calls are used only after confirming that the connection is stable. If the video fails, she moves to a phone call.

When everything else breaks down, she uses WhatsApp, either in real time or asynchronously.

This practical approach works reliably for most people, In 2021, WhatsApp was used by more than 95% of smartphone users in Nigeria, Kenya, and South Africa.[16] Princess now builds data stipends into every project and only asks for video if she truly needs it. When participants decline recordings—common when discussing money—she takes detailed manual notes instead.

Transsion Holdings (Tecno, Infinix, and Itel) accounted for 50% of smartphone shipments in Africa in 2024.[17] Most of these were entry-level Android phones priced under $100, and they drove almost all smartphone growth outside wealthy, iPhone-heavy urban areas. These devices behave differently from the mid-range and high-end phones many teams test on elsewhere. Additionally, users may share their primary phone with someone else. This shared-device context shapes how comfortable people feel testing anything related to money or identity, regardless of how "secure" the app claims to be.[18]

Network and storage issues reinforce the same pattern. As of 2024, 3G coverage still dominates at almost 90% and download speeds average 21 Mbps[19] below global averages of 50+ Mbps.[20] All of this means that if your research method requires perfect conditions, such as stable internet and private devices, you may end up studying an idealized user rather than the real one.

Regulation adds another layer of unpredictability. African regulators have tightened requirements around financial crime and KYC compliance, especially as digital payments have grown. In 2025, sub-Saharan Africa constituted 70% of global mobile money transaction value,[21] with mobile money such as M-Pesa reaching over 80% of adults in Kenya.[22] This scale attracts regulatory scrutiny—Nigeria's Central Bank issued multiple KYC and FX circulars between 2022 and 2023, tightening BVN/NIN linkages and capital controls.[23] In East Africa, transaction limits and verification requirements shift frequently. Regulatory changes can reshape user flows overnight—whether it's an extra verification step or a change in transaction limits—and enforcement differs by country.

Researchers, therefore, need to plan for uncertainty. A few approaches have proven helpful:

- Modular Flows: Build KYC and verification steps so they can be swapped in or out without breaking the whole experience. Research can then examine how users react when new requirements appear unexpectedly
- Testing Interruptions: Instead of testing only the ideal journey, test interruptions. For example, what happens if the user is halfway through sending money when they are asked to verify their ID again? Does the system explain why? Does the user know how to continue?
- Scenario Planning: Early in discovery, researchers work with product and legal teams to outline plausible regulatory shifts, such as biometric requirements and changes in daily limits, and build light or low-fidelity prototypes to test user reactions
- Being Open about Uncertainty: Sometimes the most useful approach is telling users what is still unclear. This includes testing messaging that explains unknown timelines, evolving regulations, and steps that may appear later in the process
- Treating Compliance as a Content Issue: Verification is rarely blocked by the user's technical ability. The barrier is usually understanding: Why is this required? What does "proof of address" mean in a place with no formal addressing system? Research must test the clarity of language, the tone of instructions, whether error messages help, and whether the tone feels respectful

Regulatory volatility shapes what can be built and when. But even when flows are adaptable, another factor shapes the quality of insights researchers gather: how the researcher is perceived and the power dynamics they carry into the room.

The Ethics of Positionality and Power

In many African contexts, the researcher—especially one affiliated with a global company—can easily represent privilege or institutional authority, creating an ethical tension where research itself can feel extractive if not handled with care.

People building products for African users, particularly those who are not themselves part of the communities they study, must work with a sharper awareness of positionality, power, and dignity. The YUX Design team uses "designing for dignity" as a guiding principle, rooted in simple, practical choices that reduce unnecessary power cues and help participants feel at ease.

- Hiring Local Consultants: YUX intentionally hires local researchers and cultural insiders instead of flying in external teams. Shared language, norms, and lived experience help lower barriers significantly
- Adjusting Etiquette: Teams are briefed on local expectations, such as how to greet people, when to sit, what level of eye contact is appropriate, and what clothing signals respect. Small adjustments—like greeting people respectfully or dressing simply rather than formally—can change the tone of an entire session
- Fair and Usable Compensation: Respect also shows up in how participants are paid. A bank transfer means nothing if someone is unbanked. A gift card is useless without a digital wallet. Teams must provide rewards that users can access, such as mobile airtime, data bundles, mobile money (e.g., M-PESA), or transport stipends. Crucially, in low-income communities, compensation can inadvertently lead to coercion—a sense of obligation to give positive feedback or say what the researcher wants to hear. Researchers must mitigate this by clearly communicating that honest feedback, including criticism, is more valuable than praise

Turning Research to Action

Even the richest, most careful research fails if nothing changes afterward. Across African markets, well-funded organizations commission studies, get deep, valuable insights, and then… nothing. Many international agencies, NGOs, and startups get stuck imagining only large-scale, multi-country rollouts. That ambition becomes paralyzing. Projects die not because recommendations aren't good, but because they seem too expensive, too complex, or too politically difficult to execute immediately.

YUX fights this by baking implementation thinking into every deliverable. Projects end with a prioritization matrix that sorts recommendations by impact and effort. This helps partners zero in on "high-impact, low-effort" ideas that can be piloted quickly, measured, and expanded later. They emphasize advocacy as part of the process—figuring out which gatekeepers or influencers (religious leaders, community elders, trade groups) can legitimize change within local contexts.

Small, targeted pilots prove value quickly and build trust for wider rollouts. They also respect local realities, such as testing solutions where people live and work. Every recommendation should include a pilot scope (what can be tested in one place), success metrics (how you'll know if it's working), an owner and timeline (who's leading this and when it should be delivered), and an advocacy map (which decision-makers need to be involved). But even well-planned recommendations can stall if the right people aren't convinced.

Princess learned this at Brass. Even when her research was rigorous and her insights were clear, recommendations were sometimes shelved. Leadership would override data with gut instinct. Over time, she realized that in lean, fast-moving teams, the way decisions move through power—the political side of product development—can shape outcomes as much as design itself.

This realization led her to integrate stakeholder strategy directly into the research process. Now, before running a study, she meticulously maps out who needs to be briefed, when, with what evidence, and who should be involved in synthesis. She builds credibility through consistency and visibility: sharing insights early and often and framing them in language that connects directly to business priorities. For Princess, influence is woven into the research itself; without it, even the strongest research rarely drives change.

Discovery and user research aren't phases you complete and move on from. They're practices you sustain. This means staying close to everyday signals: sitting in on customer support debriefs to hear recurring issues, using analytics and session recordings to observe real usage patterns, following up with engaged community members, speaking regularly with sales or frontline teams to understand how work happens, paying attention to social media, industry newsletters, and podcasts for qualitative signals of shifting sentiment and competition.

This kind of work is slower and less tidy than traditional research. It rarely fits into neat frameworks. But it tends to be closer to reality. When you build relationships before collecting insights, adapt methods to real conditions, and treat research as an ongoing exchange rather than a one-off exercise, the products that emerge are more likely to reflect how people live and work.

Key Takeaways

1. Traditional, global research methods (like random sampling or remote testing) may fail due to low institutional trust, unstable infrastructure, and digital literacy gaps, so teams must adapt or invent new techniques
2. To reach users outside the connected urban elite, recruitment must move to community-endorsed trust networks such as snowball sampling and local gatekeepers
3. Overcome cultural politeness and power distance by treating users as co-creators and using methods like indirect elicitation to capture genuine, sensitive user behavior

What Next?

Research Readiness Assessment

Before your next research project, complete this diagnostic to identify gaps in your approach:

Part 1: Trust Audit (score: 0–3 for each)

Rate your current research practice:

- Community Access: How do you currently recruit participants?
  - 0 = Online panels only
  - 1 = Paid recruitment services
  - 2 = Some community connections
  - 3 = Trusted gatekeepers and established relationships

- Power Dynamics: How do you address researcher privilege?
  - 0 = Haven't considered it
  - 1 = Aware but no specific actions
  - 2 = Some adjustments (dress, location)
  - 3 = Local researchers, dignity protocols, reciprocity built in
- Research Fatigue: What happens after data collection?
  - 0 = No follow-up
  - 1 = Generic thank-you
  - 2 = Share findings with participants
  - 3 = Show what changed + credit contributors

Score 7–9: Strong foundation | Score 4–6: Gaps to address | Score 0–3: Major redesign needed

Part 2: Method Reality Check

For your next planned study, ask:

1. If internet fails mid-session, what's your backup plan?
   - Write down your "modalities ladder"
2. Can participants actually access your incentive?
   - List: What you're offering/How they'll receive it/Barriers they might face
3. What direct questions might get polite deflection?
   - Identify three sensitive topics/Design indirect elicitation for each
4. Who would make participants feel MOST comfortable?
   - Map: Age/Gender/Language/Community role/Affiliation

Part 3: Implementation Planning Template

Pick ONE insight from recent research. Complete this before moving forward:

Element	Your answer
Pilot scope	What can you test in one location/segment first?
Success metric	How will you know it's working? (Be specific)
Owner and timeline	Who's leading? By when?
Advocacy map	Which gatekeepers need to approve/support this?
Barriers	What could block implementation? How will you address each?

Part 4: Weekly Discovery Rhythm (Team Exercise)

Set up your continuous listening system:

Create a shared document tracking:

- Support Themes: Top three issues from customer support (weekly)
- Behavioral Signals: Unexpected product usage patterns (from analytics)
- Community Pulse: Sentiment from social media, WhatsApp groups, forums
- Competitive Moves: What are users saying about alternatives?
- Off-Label Use: How are people repurposing features?

Action: Schedule 30-minute weekly sync to review + decide what needs deeper research

Part 5: Reflection Prompts

Individually or with your team, discuss:

1. Trust Gap: When was the last time someone didn't trust our research enough to speak honestly? What signals did we miss?
2. Method Mismatch: Which of our current research methods assume conditions that don't exist for most of our users?
3. Research Fatigue: If we've researched this community before, what did we give back? If nothing, what does that say about our values?
4. Power Blind Spots: What privilege do I carry into research sessions? How does my identity shape what people feel safe sharing?
5. Shelf Life: What percentage of our research insights from the last year actually led to product changes? If it's low, why?

Notes

1. Chui Chui Tan, *Research for Global Growth: Strategies and Guidance for Cross-Cultural Insights* (Beyō Global, 2023), 4.
2. Chui Chui Tan, *Research for Global Growth: Strategies and Guidance for Cross-Cultural Insights* (Beyō Global, 2023), 5.
3. Koffi Amessou Adaba and David Boio, *Across Africa, Public Trust in Key Institutions and Leaders Is Weakening*, Afrobarometer Dispatch No. 891 (Afrobarometer, 2024). https://www.afrobarometer.org/wp-content/uploads/2024/10/AD891-PAP20-Africans-trust-in-key-institutions-and-leaders-is-weakening-Afrobarometer-31oct24.pdf

4. Editorial Team, "Mauritius Ranks First in Africa for Digital Fraud Resilience in 2025 Global Index," *TechAfrica News*, October 29, 2025. https://techafricanews.com/2025/10/29/mauritius-ranks-first-in-africa-for-digital-fraud-resilience-in-2025-global-index/
5. Marjorie Chalwe-Mulenga, "Mitigating DFS Consumer Risks: Insights from Sub-Saharan Africa," *CGAP* (blog), August 19, 2022. https://www.cgap.org/blog/mitigating-dfs-consumer-risks-insights-sub-saharan-africa
6. Clara Delavallade et al., *Socioemotional Skills in Sub-Saharan Africa: Validating and Comparing Behavioral and Self-Reported Measures*, Policy Research Working Paper No. 11201 (World Bank, 2025). https://openknowledge.worldbank.org/entities/publication/161aa499-6b4e-46bb-a3bf-8791ba8cad14
7. Felishana Cherop et al., "'Kuchoka': Investigation of Research Fatigue in Mosoriot, Kenya," *PLoS ONE* 21, no. 1 (2025). https://pmc.ncbi.nlm.nih.gov/articles/PMC12788630/
8. Cleophas Karooma, "Research Fatigue among Rwandan Refugees in Uganda," *Forced Migration Review*, no. 61 (June 2019): 18–19. https://www.fmreview.org/ethics/karooma/
9. Rodriguez-Segura, Daniel, and Beth E. Schueler. "Assessors Influence Results: Evidence on Enumerator Effects and Educational Impact Evaluations." *Journal of Development Economics* 163 (2023): 103057. https://doi.org/10.1016/j.jdeveco.2023.103057
10. Carrie J. Ngongo et al., "The Perils of Straying from Protocol: Sampling Bias and Interviewer Effects," *PLoS ONE* 10, no. 2 (February 18, 2015): e0118025. https://pmc.ncbi.nlm.nih.gov/articles/PMC4334207/
11. CIPESA, Data Governance and Public Trust: Exploring the Sources of Low Trust Levels in Public Data Controllers in Ghana Report (April 2022), 8. https://cipesa.org/wp-content/files/documents/Data-Governance-and-Public-Trust-Exploring-the-Sources-of-Low-Trust-Levels-in-Public-Data-Controllers-in-Ghana-Report.pdf
12. Cleophas Karooma, "Research Fatigue among Rwandan Refugees in Uganda," *Forced Migration Review*, no. 61 (June 2019): 18–19. https://www.fmreview.org/ethics/karooma/
13. Morawczynski, Olga, and Gianluca Miscione. "A Longitudinal Study on M-PESA in Kenya." University of Twente, 2010, 7. https://ris.utwente.nl/ws/portalfiles/portal/326866897/MPesatrust.pdf
14. Greg Guest, Arwen Bunce, and Laura Johnson, "How Many Interviews Are Enough? An Experiment with Data Saturation and Variability," *Field Methods* 18, no. 1 (February 2006): 59–82. https://warwick.ac.uk/fac/cross_fac/q-step/currentstudents/qs103/readings/t1w3/how_many_interviews_are_enough.pdf

15. Khalifa Elmusharaf et al., "Participatory Ethnographic Evaluation and Research: Reflections on the Research Approach Used to Understand the Complexity of Maternal Health Issues in South Sudan," *Qualitative Health Research* 27, no. 9 (2017): 1345–58. https://pmc.ncbi.nlm.nih.gov/articles/PMC5476190/
16. Verint Systems, "What Countries Are the Biggest WhatsApp Users?" *Verint* (blog), December 22, 2021. https://www.verint.com/blog/what-countries-are-the-biggest-whatsapp-users/
17. Yicai Global, "Star 50 News, February 27, 2025." https://www.yicaiglobal.com/star50news/2025_02_276798560138878779401
18. See Chapter 3, "Designing Mobile-First for African Contexts," and Chapter 4, "Building for Digital Connectivity."
19. "The State of Mobile Internet Connectivity 2025", GSMA, accessed August 30, 2025 https://www.gsma.com/somic/
20. Best Broadband Deals, "Worldwide Speed League," Best Broadband Deals. Accessed November 25, 2025. https://bestbroadbanddeals.co.uk/broadband/speed/worldwide-speed-league/#regions
21. Africa's Payments Revolution: Leading the Global Mobile Money Boom." Institute for Inclusive Digital Africa, April 4, 2025. https://www.iidia.org/africas-payments-revolution-leading-the-global-mobile-money-boom/
22. Central Bank of Kenya, Kenya National Bureau of Statistics, and FSD Kenya, 2024 *FinAccess Household Survey* (Central Bank of Kenya, 2024), 12. https://www.centralbank.go.ke/wp-content/uploads/2024/12/2024-FINACCESS-HOUSEHOLD-SURVEY-MAIN-REPORT.pdf
23. Central Bank of Nigeria (CBN), *CBN Customer Due Diligence Regulations*, 2023 (CBN, 2023). https://www.cbn.gov.ng/Out/2023/CCD/CBN%20Customer%20Due%20diligence%20Reg.%202023-combined.pdf

Bibliography

Adaba, Koffi Amessou, and David Boio. *Across Africa, Public Trust in Key Institutions and Leaders Is Weakening.* Afrobarometer Dispatch No. 891. Afrobarometer, 2024. https://www.afrobarometer.org/wp-content/uploads/2024/10/AD891-PAP20-Africans-trust-in-key-institutions-and-leaeders-is-weakening-Afrobarometer-31oct24.pdf

Central Bank of Kenya, Kenya National Bureau of Statistics, and FSD Kenya. *2024 FinAccess Household Survey Report.* Central Bank of Kenya, 2024. https://www.centralbank.go.ke/uploads/finaccess/1681216140_FinAccess_2024_Report.pdf

Central Bank of Nigeria. *CBN Customer Due Diligence Regulations, 2023.* CBN, A2023. https://www.cbn.gov.ng/Out/2023/CCD/CBN%20Customer%20Due%20diligence%20Reg.%202023-combined.pdf.

Chalwe-Mulenga, Marjorie. "Mitigating DFS Consumer Risks: Insights from Sub-Saharan Africa." *CGAP* (blog). August 19, 2022. https://www.cgap.org/blog/mitigating-dfs-consumer-risks-insights-sub-saharan-africa.

Cherop, Felishana, Violet Naanyu, Juddy Wachira, and Lukoye Atwoli. "'Kuchoka': Investigation of Research Fatigue in Mosoriot, Kenya." *PLoS ONE* 21, no. 1 (2025). https://pmc.ncbi.nlm.nih.gov/articles/PMC12788630/

CIPESA. Data Governance and Public Trust: Exploring the Sources of Low Trust Levels in Public Data Controllers in Ghana Report. April 2022. https://cipesa.org/wp-content/files/documents/Data-Governance-and-Public-Trust-Exploring-the-Sources-of-Low-Trust-Levels-in-Public-Data-Controllers-in-Ghana-Report.pdf.

Editorial Team. "Mauritius Ranks First in Africa for Digital Fraud Resilience in 2025 Global Index." *TechAfrica News*, October 29, 2025. https://techafricanews.com/2025/10/29/mauritius-ranks-first-in-africa-for-digital-fraud-resilience-in-2025-global-index/

Elmusharaf, Khalifa, Elaine Byrne, Mary Manandhar, Joanne Hemmings, Diarmuid O'Donovan. "Participatory Ethnographic Evaluation and Research: Reflections on the Research Approach Used to Understand the Complexity of Maternal Health Issues in South Sudan." *Qualitative Health Research* 27, no. 9 (2017): 1345–58. https://pmc.ncbi.nlm.nih.gov/articles/PMC5476190/

GSMA. "The State of Mobile Internet Connectivity 2025". Accessed August 30, 2025. https://www.gsma.com/r/somic/.

Guest, Greg, Arwen Bunce, and Laura Johnson. "How Many Interviews Are Enough? An Experiment with Data Saturation and Variability." *Field Methods* 18, no. 1 (February 2006): 59–82. https://warwick.ac.uk/fac/cross_fac/q-step/currentstudents/qs103/readings/t1w3/how_many_interviews_are_enough.pdf

Institute for Inclusive Digital Africa. "Africa's Payments Revolution: Leading the Global Mobile Money Boom." April 4, 2025. https://www.iidia.org/africas-payments-revolution-leading-the-global-mobile-money-boom/

ITU and UNESCO. *The State of Broadband in Africa 2025*. Broadband Commission, 2025. https://www.broadbandcommission.org/wp-content/uploads/dlm_uploads/2025/09/The-State-of-Broadband-in-Africa.pdf

Karooma, Cleophas. "Research Fatigue among Rwandan Refugees in Uganda." *Forced Migration Review*, no. 61 (June 2019): 18–19. https://www.fmreview.org/ethics/karooma/

Morawczynski, Olga, and Gianluca Miscione. "A Longitudinal Study on M-PESA in Kenya." Enschede: University of Twente, 2010. https://ris.utwente.nl/ws/portalfiles/portal/326866897/MPesatrust.pdf

Ngongo, Carrie J., Kevin D. Frick, Allen W. Hightower, Florence Alice Mathingau, Heather Burke, and Robert F. Breiman. “The Perils of Straying from Protocol: Sampling Bias and Interviewer Effects.” *PLoS ONE* 10, no. 2 (February 18, 2015): e0118025. https://pmc.ncbi.nlm.nih.gov/articles/PMC4334207/

Rodriguez-Segura, Daniel, and Beth E. Schueler. "Assessors influence results: Evidence on enumerator effects and educational impact evaluations." *Journal of Development Economics* 163 (2023): 103057. https://doi.org/10.1016/j.jdeveco.2023.103057

Tan, Chui Chui. *Research for Global Growth: Strategies and Guidance for Cross-Cultural Insights*. Beyō Global, 2023.

Verint Systems. “What Countries Are the Biggest WhatsApp Users?” *Verint* (blog). December 22, 2021. https://www.verint.com/blog/what-countries-are-the-biggest-whatsapp-users/.[edelman]

Yicai Global. “The competition in the African market is fierce, and Transsion Holdings’ gross profit margin and non-net profit will both decline in 2024.” February 27, 2025. https://www.yicaiglobal.com/star50news/2025_02_276798560138878779401

Chapter 8
The Future of Design in Africa

In the early 2020s, cryptocurrency was taking hold across Africa. Its appeal stemmed from its infrastructure—smart contracts acting as trusted, code-based intermediaries and blockchain as a public ledger—offering a promise of borderless, resilient finance in regions with volatile economies and high mobile usage. Seeing this opportunity in 2021, at the height of the cryptocurrency bull market, Emmanuel Njoku, a 19-year-old blockchain engineer from Nigeria who had previously worked for companies like MakerDAO, was determined to build what he described as the "Paystack of crypto."[1] He founded Lazerpay—a smart-contract-enabled payments gateway that allowed merchants to accept cryptocurrency, including stablecoins. The platform quickly gained traction, showcasing the strong market demand for an alternative financial rail.

By Q3 2022, Africa had become "the fastest-growing cryptocurrency market among developing economies and the third-fastest-growing market globally."[2] However, the collapse of FTX in November 2022 marked a turning point. It intensified the broader crypto downturn and triggered a crisis of trust across the industry. By April 2023, Lazerpay had shut down.[3]

DOI: 10.1201/9781003648192-11

While tighter crypto regulations in Nigeria contributed to this outcome, some argue the company was simply ahead of its time. Despite the global "crypto winter" of 2023, adoption continued to rise in sub-Saharan Africa. Nigeria alone accounted for $59 billion in on-chain value, with South Africa following at over $25 billion.[4] In 2024, retail stablecoin transactions in South Africa and Uganda increased by 50% compared to the previous year.[5] This growth has been driven by everyday use—particularly as high fees and friction in traditional cross-border payments push people toward alternatives such as stablecoins.

Stories like Lazerpay's are often discussed through the lens of innovation or failure. Yet they are just as revealing from a design perspective. Building digital products in African contexts has rarely been about technical novelty alone. It has required designing for instability, fragmented regulation, uneven infrastructure, and users who rely on products not as conveniences but as essential tools for navigating everyday life. Their story, while painful, powerfully illustrates how a purely product-focused mindset often fails to account for the full service ecosystem—the combination of regulatory, financial, and logistical rails that must be designed for resilience.

In these environments, success is determined less by how advanced a technology is and more by whether it fits into existing systems, behaviors, and workarounds. As Africa continues to adopt new technologies earlier and at scale, the future of design on the continent will increasingly be shaped by this reality. Those building digital experiences will need to think beyond interfaces and features and instead consider entire services: how people onboard, how trust is built, how failures are handled, and how products coexist with informal practices.

From Design Thinking to Service Thinking

As product researchers, designers, managers, builders, or problem solvers in general, one of the first methodologies we encounter is design thinking. Design thinking is widely described as a flexible, repeatable, and human-centered approach to creative

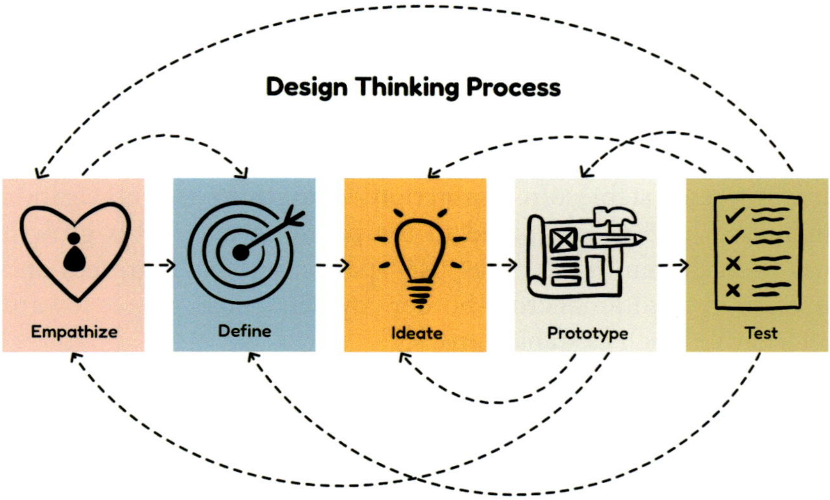

FIGURE 8.1 The design thinking process.

problem-solving. Rather than jumping straight into development, it prioritizes empathy—the ability to deeply understand the people, environments, and challenges you are designing for.

In *The Design Thinking Toolbox*, the authors introduce the concept of the "beginner's mind" as a core foundation of design thinking.[6] They describe it as a form of deep curiosity—an openness to learning how things work and why they work the way they do. This kind of curiosity, which we have explored in earlier chapters, is especially important in environments where assumptions travel poorly. This beginner's mind—a refusal to accept things as simply "the way they are"—is what Yanick Kemayou sees thriving in African communities[7].

Kemayou, co-founder of Kabakoo Academies—an edtech and innovation lab working across the Sahel—argues that while design thinking is often framed through Western contexts, its principles are alive in how people across the continent solve problems.[8] Long before it was formalized into workshops and toolkits, the core ideas behind design thinking—non-linear problem-solving, iteration, and solutions grounded in real-world contexts—were already present in how people navigated constraints. This ingenuity is visible in the workarounds users develop to make products usable for their needs, a phenomenon discussed further in Chapter 3. It is also evident in Kabakoo's work, which now includes an AI-powered learning tool and has helped 42% of its learners secure training-related opportunities within six months.[9]

Design thinking shows up in African contexts in a few consistent ways. Empathy often takes the form of participatory research that actively reduces power dynamics—through how researchers show up, how they dress, where conversations happen, and whose behaviors set the tone. Problem definition is rarely abstract, as it requires understanding the workarounds people already rely on and reframing the problem in light of those realities. Ideation is shaped by constraints, producing solutions that account for infrastructure gaps, economic pressure, and risk sensitivity. Prototyping and testing tend to be rapid and continuous, often happening through customer support calls, WhatsApp conversations, or repeated confirmation steps designed to reduce anxiety around money and trust.

Taken together, these examples point to a clear shift. While design thinking remains a useful foundation, its future in African contexts lies in Service Thinking: designing not just products, but the full ecosystem around them, shaped by trust, culture, constraint, and lived behavior. This approach is already visible in large, service-heavy organizations such as public services and traditional banks in the UK and Europe, where service design is used to make complex systems clearer and more navigable. At the core of Service Design is Service Thinking.

Consider this real scenario: A fintech company in Kenya dedicated a year to perfecting its loan application interface using design thinking. The flow was clean, and the copy was clear. However, by narrowly applying design thinking to the feature and not the full end-to-end service, they missed crucial points. After launch, approval rates were terrible, and customer complaints skyrocketed because they hadn't designed for:

- What happened when users needed to upload documents but had poor connectivity (the upload would fail silently, and users thought their application was processing)
- How credit officers on the backend reviewed applications (they needed information the form didn't collect)

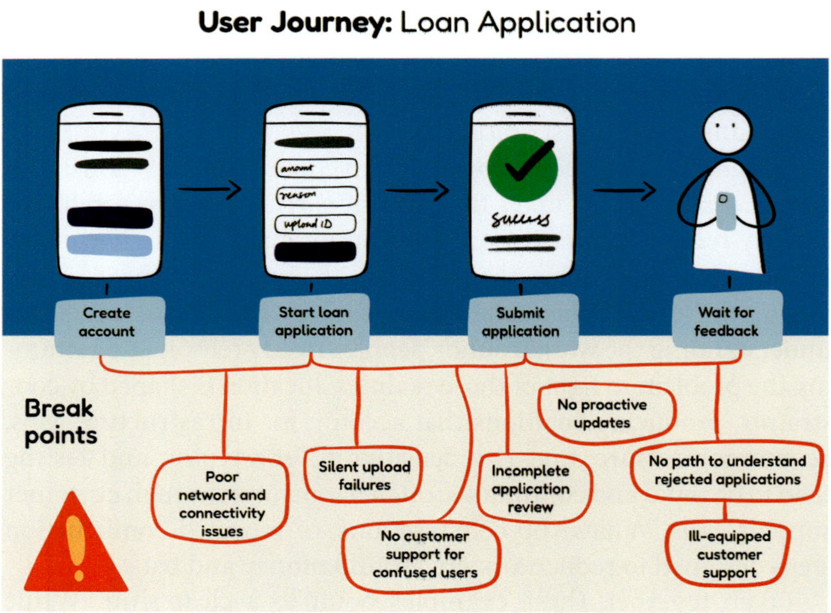

FIGURE 8.2 User journey highlighting break points because of an absence of Service Thinking.

- How users would know their application status without expensive phone calls (no proactive SMS updates)
- What users would do when rejected without explanation (no path to understand why or improve their application)
- How customer support would handle the volume of confused users (support wasn't equipped with the right information)

This shortfall illustrates the difference between applying design thinking to a single feature and applying it to the entire service ecosystem—which is where Service Thinking becomes essential. If design thinking asks, "How do we make this interaction work well?", Service Thinking expands the scope to ask, "How do we make this entire service work well, even when things go wrong?" The practical difference in scope is clear and highlighted in Table 8.1.

This ideology aligns directly with the African Product Resilience (APR) Framework™. Service Thinking requires Infrastructure Agnosticism—you cannot design a complete service without

TABLE 8.1 Comparative Inquiries: Design Thinking versus Service Thinking

Design Thinking Asks:	Service Thinking Asks:
Who is the user and what do they need right now?	Who are all the actors in this ecosystem (users, agents, support staff, regulators), and how do they interact?
How do we make this feature intuitive?	How does this feature connect to everything before and after it in the user's journey?
What does the user see and experience?	What needs to happen behind the scenes—operationally, technically, organizationally—to deliver on what the user expects?
How do we solve this problem elegantly?	How do we build organizational capabilities to keep solving problems as they evolve?

accounting for connectivity failures, power outages, and device limitations. It also demands trust-centered interactions—because trust is built not in a single transaction but across an entire service relationship, including how failures are handled and how users are treated when they need help.

WHAT GOOD SERVICE THINKING LOOKS LIKE

In *Good Services*, Lou Downe, former Design Director of the UK Government, outlines principles for good service design that—despite emerging from a different context—speak directly to the challenges African products face.[10] Downe's principles, forged through making government services work for millions of people, including marginalized and vulnerable populations, resonate because they address the same fundamental challenge: how do you design services that work for everyone, including when things go wrong? According to Downe, under Service Thinking, a good product or service must[11]:

- Be Easy to Discover and Access: This means being findable through the channels people use, such as word of mouth, not just app stores
- Clearly Communicate What It's for and When to Use It: This prevents users from attempting tasks your service doesn't support, which wastes their time, data, and trust.

If your service is for bill payments but users think it's for money transfers, they'll have a failed experience before they even understand what went wrong

- Set Clear Expectations about How It Works and What Users Can Expect: This is especially important for financial services, where unexpected behavior can instantly erode trust
- Help Users Complete the Task and Outcome They Set Out to Do: This means designing for task completion when users need to switch devices or pause and return later
- Fit Naturally into Behaviors People Already Understand: This is why M-Pesa's agent model worked—it fit into existing patterns of money transfer through human intermediaries
- Require No Prior Knowledge to Use: Essential when users may be encountering your service category for the first time
- Be Agnostic of Internal Team or Company Structure: Users shouldn't have to know that payments and customer service are different departments; the service should feel unified
- Avoid Dead Ends—Always Offer a Next Step or Recovery Path: When something fails, the service itself must guide users to resolution
- Be Usable by Everyone: This includes people with low literacy, people sharing devices, people with inconsistent connectivity, and people using budget smartphones
- Adapt Quickly as Needs, Conditions, or Contexts Change: Services must be designed to evolve as infrastructure improves, as user sophistication increases, and as regulatory environments shift
- Make It Easy to Reach a Human When Automation Falls Short: Through channels users can afford and access, whether that's WhatsApp, callback requests, or agent networks

These principles are practical requirements for products and services that must work reliably in complex, constraint-heavy environments. They push product builders to think beyond the screen to the entire support system that makes a service function in users' lives.

Service Thinking in Practice

If the Kenyan fintech mentioned earlier used Service Thinking from the start, they would have:

- Mapped the entire journey from awareness to loan disbursement to repayment, identifying every point where users might get stuck
- Designed clear status communication that worked via SMS, not just in-app notifications users might not see
- Built the application form based on what credit agents needed to make decisions, not just what seemed logical from a user perspective
- Created explicit failure paths with clear next steps when documents failed to upload or applications were rejected
- Equipped customer support with the information and tools to resolve issues, not just apologize
- Tested not just the happy path but all the ways things could go wrong—and designed for those scenarios

This is more work upfront, as it requires coordinating across teams—product, operations, support, and compliance. For startups with limited resources, this might sound overwhelming, but designing only the interface and leaving everything else unplanned doesn't save time or money—it just moves the work to after launch, when it's more expensive to fix and you're bleeding users who've had bad experiences.

Service Thinking forces teams to design for the whole system, not just the interface. In African contexts, this leads to more resilient products that can handle real-world complexity. As product teams begin to grasp this shift from isolated features to complete service ecosystems, another force reshaping what's possible is artificial intelligence.

AI as an Amplifier of Services

The estimated market size for AI in Africa in 2025 is \$4.51 billion and is expected to reach \$16.53 billion by 2030, driven by speech recognition, image processing, autonomous vehicles, and services that enable organizations to develop and deploy AI applications.[12]

Like many waves of technological change on the continent, AI adoption is not moving evenly. Momentum is concentrated in fintech, healthcare, education, and agriculture, where companies are experimenting with AI-driven tools—largely in a handful of cities among the big four markets and among digitally fluent users. Outside these hubs, AI is far less visible, shaped by the same constraints that have always defined digital product adoption in Africa: infrastructure and access.

Africa holds less than 1% of global data center capacity, even though data centers are the engines that power modern AI.[13] Cloud computing infrastructure is sparse and uneven, which means most African startups rely on foreign providers like AWS or Google Cloud. This dependence introduces trade-offs: higher costs, slower performance due to distance from data centers, and uncertainty around cross-border data transfer and regulation.

Another challenge is that AI systems deployed on the continent are trained on datasets built elsewhere. Only 2% of data used to train global AI models comes from Africa.[14] For a continent with over 2,000 languages, distinct transaction patterns, cultural practices, and contexts that may not exist anywhere else, this is problematic.

When fraud detection algorithms learn what "suspicious" looks like from American credit card data, they flag normal Nigerian mobile money patterns as fraudulent. When voice recognition systems are trained on American English, they struggle when someone code-switches between Yoruba and Pidgin or speaks Kiswahili with a Kenyan accent in a noisy area. The AI works technically, but it has not yet been built for the contexts where it's being deployed. This is the same pattern we've seen throughout this book: technology designed elsewhere, for other assumptions, now expected to work in environments where those assumptions don't hold.

Where AI Is Being Used

Despite these infrastructure constraints, AI adoption is happening in specific sectors where the value proposition is clear enough to justify working around the limitations.

By sector, fintech leads by a significant margin. AI powers customer service chatbots that handle routine queries, fraud detection systems that flag suspicious transactions, and credit scoring algorithms that assess loan eligibility. The latter application is important because traditional credit histories don't exist for most people. In Kenya, Tala assesses creditworthiness by analyzing data from mobile phones including travel patterns and payment behavior.[15] It is extending credit to people banks would automatically reject, though it raises uncomfortable questions about privacy and what data should determine financial access—questions we'll return to shortly.

Agriculture is another central frontier for AI innovation in Kenya, South Africa, Nigeria, and Egypt, with companies applying the technology to solve deeply practical, real-world challenges.[16] Farmerline connects farmers to market prices, quality inputs, and agronomic advice through Darli AI—accessible via WhatsApp chatbots and Interactive Voice Response in over 27 languages, including Twi, Swahili, and Yoruba.[17] A farmer can ask about crop diseases, check market prices, and get weather forecasts, all from their phone in their language.

In education, AI is enabling a kind of personalization that would be impossible through traditional teaching at scale. Rising Academies' Rori is an AI-powered math tutor on WhatsApp that adapts to each student's learning pace. It's currently working with low-literacy, low-income students across Sierra Leone and Ghana,[18] helping them build foundational math skills through conversations in natural language.

These are tools solving real problems. However, there's a pattern in these examples and many others like them: success is concentrated in specific use cases where the technology can work within existing constraints, not where it requires perfect conditions that don't exist for most users. The fintech chatbots work because text-based interaction doesn't require much bandwidth, and the education app works because it can cache lessons and work partially offline.

Voice Interfaces: Promise and Practical Limits

Among all AI applications for African contexts, voice technology might be the most promising. At the same time, it clearly shows the gap between what the technology could enable and what current infrastructure and training data can realistically support.

For users with low literacy, voice interfaces could unlock services that text-based designs never will. Yann Le Beux, whose research agency YUX has spent years working across seven African countries, describes voice-to-text and text-to-speech as "almost there"—they work reasonably well when speakers have clear pronunciation in languages like Wolof or Yoruba and when there's minimal background noise.

The problem is that in real-world African contexts, you almost never have both conditions at once. Think about where people use their phones. Markets, for example, are noisy with vendors calling out prices and music playing. Street corners often have traffic and multiple conversations happening simultaneously. In addition, people don't always speak in the neat, single-language sentences that voice AI expects—as code-switching between languages mid-sentence is how millions of people in this part of the world communicate every day.

So voice interfaces designed for quiet environments with single-language input fail in these conditions. They misunderstand and transcribe gibberish, leaving users frustrated and feeling like the technology isn't "for them." Which, to be honest, it isn't—at least not yet. This calls for more thoughtfulness in how we deploy voice.

YUX Design is working on the harder, longer-term problem: training speech recognition models on how people talk in African contexts. Not "English with a slight accent filter," as Le Beux puts it, but models that genuinely understand code-switching, colloquialisms, multiple speakers talking over each other, background noise—all the messy reality of real communication. This requires collecting training data from actual environments. It's crucial work because it addresses the fundamental mismatch between how AI has been trained and how people communicate.

What Gets Built versus What's Needed

Building AI that genuinely fits African realities requires different choices: training voice models with code-switching from the start, designing for intermittent connectivity, developing credit algorithms based on African transaction data, and building agricultural tools using local weather patterns and crops. This work depends on local data collection, research capacity, talent development, and investment in infrastructure such as data centers.

Some of this is starting to happen. Universities are launching AI programs—OpenAI recently selected the University of Lagos to host its first AI academy in Africa.[19] Startups are also building locally relevant models with patient capital from investors who understand the long game.

However, AI does not solve infrastructure problems—it often makes them more visible by demanding more computing power and connectivity. It also does not remove the need for human judgment or local knowledge. And it does not benefit everyone equally. AI systems trained on limited or non-representative data serve some users far better than others.

For AI to expand access rather than reinforce existing inequalities, it must be built with realistic assumptions about infrastructure, careful attention to data costs and offline use, transparency in decision-making, and recognition that technology alone cannot solve social problems. The tools are becoming more capable. Whether they help address real user needs or primarily benefit those who already have the most resources depends on the choices made in how they are designed and deployed.

Areas of Active Research

Several technical approaches are being explored to make AI more viable in African contexts, though most remain in research or early pilot stages.

- On-Device AI Processing: Running AI models locally on phones rather than in the cloud could reduce data costs and dependence on connectivity. Tech companies like Google and Meta are developing smaller models that can work on mobile devices, though they haven't been optimized for African languages or contexts yet
- Federated Learning: This approach trains AI models across many devices without centralizing data, addressing data

privacy concerns. However, poor internet connectivity and the technical complexities associated with this method could present constraints[20]

- Lightweight AI Architectures: Researchers are exploring how to build AI systems that require less computational power and data, making them more feasible in resource-constrained environments. This work is happening in academic institutions across the continent, though commercial applications remain limited

These approaches address actual constraints: expensive data, unreliable connectivity, and limited computational infrastructure. However, how quickly they'll move from research to deployed products accessible to ordinary users remains uncertain and will largely depend on whether investment and policy support can sustain local innovation long enough for it to mature.

The Next Frontier: Emerging Threads

The foundations we have explored in this book are the launchpad for the next decade of African digital innovation. Looking forward, we see several interwoven threads that will shape design practice across the continent:

- Voice and Multimodal Interaction as the Standard: Products will move beyond a single text-based interface, assuming users need to speak, type, or tap seamlessly. Although this requires sustained investment in robust African language datasets, this shift will significantly lower the barrier for users with limited literacy
- The Rise of Hybrid Physical-Digital Systems: Purely digital solutions will increasingly be replaced by powerful hybrid experiences. The most successful products will integrate agent networks, physical distribution, robust offline support, and digital interfaces into one coherent system. The boundary between a "digital product" and a "service with

digital components" will effectively disappear, reflecting the reality we see with M-Pesa and Moniepoint

- Trust-Building Remains the Core Currency: As platforms scale and digital threats evolve, the products that win will be the ones users can count on completely. Trust isn't something you just add to a product—it will be woven into every interaction and every moment when something goes wrong. Similarly, design's job extends beyond making tasks easier to making the entire system feel believable and safe
- Decentering Design with Participatory Methods: The shift toward local, context-driven research will move into the product build itself. Participatory design methods, where communities co-create solutions rather than just being studied, will become the norm. Research will look less like extraction and more like genuine collaboration, resulting in products that communities want to keep using
- African Design as a Global Blueprint: The principles of the African Product Resilience (APR) Framework™—resilient architecture, constraint-driven creativity, trust-centered UX—are already influencing global practice. As African builders gain greater visibility, the flow of design influence will become bidirectional. The world will learn from proven models for building technology that works for everyone, not just the well-connected
- Unlocking Possibilities through Infrastructure: Slow but steady improvements in reliable power, faster connectivity, and cheaper devices will remove constraints and open up new design possibilities. The real challenge will be making sure these investments reach beyond urban centers into the communities that need them most

The Usability Bible's Final Word: A Future of Resilience

The future of African design, as we have charted throughout this book, is not a diluted version of a Western-centric vision. It will be shaped by builders who understand that the most transformative innovation is often born from the greatest constraints, constraints that turn adversity into a competitive advantage.

That future is already being built. It's being built by designers like Oniyitan, who obsess over making every tap count and over localization that goes deeper than translation. It's being built by researchers like Audu, who involve communities as collaborators and advocate for ethical AI and inclusive datasets. It's being built by product managers like Akari, who refuse to ship without understanding context deeply.

It's being built by product veterans like Adeniran who know that infrastructure comes before interface and that the most transformative innovations often look boring until you understand what they unlock.

The future of African design is human-centered in the deepest sense: grounded in how people actually live and built on constraints that demand creativity. It's about building products that fit African reality—and in the process, teaching the rest of the world what resilient, inclusive, trust-centered design looks like.

Key Takeaways

1. While Design Thinking is a useful foundation, the future in African contexts requires Service Thinking: expanding the design scope from a single feature to the entire service ecosystem—including operations, support, failure paths, and all human actors—to ensure resilience when things go wrong
2. The African AI market is growing rapidly, with momentum concentrated in specific sectors. However, AI adoption is severely constrained by sparse data center infrastructure and the use of global models trained on non-African data, leading to a mismatch between technology and local context
3. Voice interfaces hold great promise for users with low literacy, but they consistently fail in the context of noisy environments, code-switching, and poor connectivity. Simple, locally cached solutions like recorded audio explanations are often more effective and resilient than sophisticated, real-time NLP

4. The principles of the African Product Resilience (APR) Framework™—resilience, constraint-driven creativity, and trust-centered UX—will increasingly influence global practice, providing the world with proven models for building universal and inclusive technology

What Next?

Resilience and Service Thinking Audit

System and Service Flow

- Have we mapped the full end-to-end user journey, including all operational and human touchpoints (e.g., agents, support staff)?
- Have we designed explicit failure paths with clear next steps for users when the experience breaks (e.g., connectivity loss, payment failure, rejection)?
- Does the service integrate non-digital components, like agent networks or physical distribution, into a coherent hybrid system?
- Does our communication channel plan (e.g., SMS, WhatsApp, in-app) ensure users receive critical status updates even when they are offline?

System and Service Flow

- Is the product functional and usable under intermittent connectivity (2G/3G) and with low-power battery anxiety?
- Have we optimized code and data transfer to keep the app lightweight and minimize data costs for the user?
- Does the service support a clear, low-data method for users to reach a human support channel when automation fails?

AI and Data Context Audit

Model and Training Data

- Is our AI system trained on datasets that represent the diversity of African languages, accents, code-switching patterns, and behaviors?

- Have we tested the AI model's performance on budget devices with limited on-device processing capabilities?
- If using third-party AI, what are the performance and cost trade-offs of relying on distant cloud infrastructure versus developing lightweight, local models?

Voice and Multimodal Design

- For low-literacy users, have we prioritized resilient audio solutions (like cached recorded audio) over real-time Natural Language Processing (NLP) that may fail in noisy environments?
- Do we support multimodal input (speak, type, tap) fluidly, ensuring one failure point doesn't block the entire task?

Trust and Accountability Audit

Transparency and Control

- Is the data collection and usage policy transparent, especially for sensitive areas like credit scoring or fraud detection?
- Are users given control over their data, aligning with rights such as consent and erasure?
- Is it clear to the user when they are interacting with AI/automation versus a human agent?

Notes

1. "Njoku Emmanuel's Lessons from a $1 Million Startup Failure." Posted August 15, 2024, by "The Flip," YouTube, 6:04. https://www.youtube.com/watch?v=EqKIsmzi7rs
2. Triple-A, "Cryptocurrency Ownership Data: Nigeria." Accessed September 11, 2025. https://www.triple-a.io/cryptocurrency-data/nigeria
3. "Njoku Emmanuel's Lessons from a $1 Million Startup Failure." Posted August 15, 2024, by "The Flip," YouTube, 2:30. https://www.youtube.com/watch?v=EqKIsmzi7rs
4. Chainalysis, "Sub-Saharan Africa: Nigeria Takes #2, South Africa Grows Crypto Adoption," October 2, 2024. https://www.chainalysis.com/blog/subsaharan-africa-crypto-adoption-2024/
5. Chainalysis, "Sub-Saharan Africa: Nigeria Takes #2, South Africa Grows Crypto Adoption," October 2, 2024. https://www.chainalysis.com/blog/subsaharan-africa-crypto-adoption-2024/

6. Michael Lewrick, Patrick Link, and Larry Leifer, T*he Design Thinking Toolbox: A Guide to Mastering the Most Popular and Valuable Innovation Methods* (Wiley, 2020), 18.
7. Yanick Kemayou, "Rethinking Design Thinking for Africa," *Kabakoo* (Medium), August 25, 2019. https://medium.com/kabakoo/rethinking-design-thinking-for-africa-5d69cf719fb1
8. Yanick Kemayou, "Rethinking Design Thinking for Africa," *Kabakoo* (Medium), August 25, 2019, para. 4–6. https://medium.com/kabakoo/rethinking-design-thinking-for-africa-5d69cf719fb1
9. Kabakoo Academies, "Digital Upskilling," accessed November 7, 2025. https://www.kabakoo.africa/digital-upskilling
10. Lou Downe, *Good Services: How to Design Services That Work* (BIS Publishers, 2019).
11. Lou Downe, *Good Services: How to Design Services That Work* (BIS Publishers, 2019) 45.
12. Mastercard, *Harnessing the Transformative Power of AI in Africa* (Mastercard, 2025), 3. https://www.mastercard.com/content/dam/mccom/shared/news-and-trends/insights/2025/ai-and-africa/mastercard-ai-in-africa-2025.pdf
13. Africa Data Centres Association (ADCA) and Rising Advisory, *Data Centres in Africa 2026* (ADCA 2026), 6. https://africadca.org/wp-content/uploads/2026/02/DataCentresInAfrica2026-EN.pdf
14. Deutsche Gesellschaft für Internationale Zusammenarbeit (GIZ), "Africa's AI Revolution Needs You: Let's Build It With Data," January 10, 2025. https://www.giz.de/en/regions/africa/african-union/news/africas-ai-revolution-needs-you-lets-build-it-data
15. Rachel Layne, "How $40 Loans Lifted Lives in Kenya," Working Knowledge, April 24, 2025. https://www.library.hbs.edu/working-knowledge/how-40-dollar-loans-lifted-lives-in-kenya
16. International Center of Expertise in Montreal on Artificial Intelligence (CEIMIA), *State of AI in Agriculture in Sub-Saharan Africa* (CEIMIA, 2024). https://ceimia.org/wp-content/uploads/2024/07/state-of-ai-in-agriculture-sub-saharan-africa_25-07-2024-docx.pdf
17. Jeff Wilser, "Farmerline Darli AI: A Chatbot for Farmers," *TIME*, October 30, 2024. https://time.com/7094874/farmerline-darli-ai
18. "About Rori", Rori AI. Accessed February 12, 2026. https://rori.ai/about
19. Mariam Ileyemi, "UNILAG Partners OpenAI to Launch Africa's First AI Academy," *Premium Times*, October 15, 2025. https://www.premiumtimesng.com/news/top-news/828150-unilag-partners-openai-to-launch-africas-first-ai-academy.html
20. Jorge Fabila et al., "Federated Learning in Low-Resource Settings: A Chest Imaging Study in Africa—Challenges and Lessons Learned," *arXiv*:2505.14217v1 [cs.LG], May 20, 2025. https://doi.org/10.48550/arXiv.2505.14217

Bibliography

Africa Data Centres Association and Rising Advisory. *Data Centres in Africa 2026*. ADCA, 2026. https://africadca.org/wp-content/uploads/2026/02/DataCentresInAfrica2026-EN.pdf

Chainalysis. "Sub-Saharan Africa: Nigeria Takes #2, South Africa Grows Crypto Adoption." October 2, 2024. https://www.chainalysis.com/blog/subsaharan-africa-crypto-adoption-2024/

Deutsche Gesellschaft für Internationale Zusammenarbeit (GIZ). "Africa's AI Revolution Needs You: Let's Build It With Data," January 10, 2025. https://www.giz.de/en/regions/africa/african-union/news/africas-ai-revolution-needs-you-lets-build-it-data

Downe, Lou. *Good Services: How to Design Services That Work*. BIS Publishers, 2019.

Fabila, Jorge, Lidia Garrucho, Víctor M. Campello, Carlos Martín-Isla, and Karim Lekadir. "Federated Learning in Low-Resource Settings: A Chest Imaging Study in Africa—Challenges and Lessons Learned." *arXiv*:2505.14217v1 [cs.LG], May 20, 2025. Accessed February 26, 2026. https://doi.org/10.48550/arXiv.2505.14217

Ileyemi, Mariam. "UNILAG Partners OpenAI to Launch Africa's First AI Academy." *Premium Times*, October 14, 2025. https://www.premiumtimesng.com/news/top-news/828150-unilag-partners-openai-to-launch-africas-first-ai-academy.html

International Center of Expertise in Montreal on Artificial Intelligence (CEIMIA). *State of AI in Agriculture in Sub-Saharan Africa*. CEIMIA, 2024. https://ceimia.org/wp-content/uploads/2024/07/state-of-ai-in-agriculture-sub-saharan-africa_25-07-2024-docx.pdf

Kabakoo Academies. "Digital Upskilling." Accessed November 7, 2025. https://www.kabakoo.africa/digital-upskilling

Kemayou, Yanick. "Rethinking Design Thinking for Africa." *Kabakoo* (Medium). August 25, 2019. https://medium.com/kabakoo/rethinking-design-thinking-for-africa-5d69cf719fb1

Layne, Rachel. "How $40 Loans Lifted Lives in Kenya." Working Knowledge, April 24, 2025. https://www.library.hbs.edu/working-knowledge/how-40-dollar-loans-lifted-lives-in-kenya

Lewrick, Michael, Patrick Link, and Larry Leifer. *The Design Thinking Toolbox: A Guide to Mastering the Most Popular and Valuable Innovation Methods*. Wiley, 2020.

Mastercard. *Harnessing the Transformative Power of AI in Africa*. Mastercard, 2025. https://www.mastercard.com/content/dam/mccom/shared/news-and-trends/insights/2025/ai-and-africa/mastercard-ai-in-africa-2025.pdf

"Njoku Emmanuel's Lessons from a $1 Million Startup Failure." Posted August 15, 2024, by The Flip. YouTube, 6:04. https://www.youtube.com/watch?v=EqKIsmzi7rs

Rori AI. "About Rori." Accessed February 12, 2026. https://rori.ai/about

Triple-A. "Cryptocurrency Ownership Data: Nigeria." Accessed November 7, 2025. https://www.triple-a.io/cryptocurrency-data/nigeria

Wilser, Jeff. "Farmerline Darli AI: A Chatbot for Farmers." *TIME*, October 30, 2024. https://time.com/collections/best-inventions-2024/7094874/farmerline-darli-ai/

Appendix: Voices of African Product Resilience™

In this section, we have put together a collection of case studies and written contributions gathered from experts across the continent. These include founders, product managers, designers, software engineers, researchers, and marketing professionals in the tech industry, who share insights from their work on the following themes:

1. Building for Digital Access
2. Building for Digital Literacy
3. Building for Socio-Cultural Context
4. Building for Trust
5. Conducting User Research in Africa
6. Navigating Regulations in the African Context
7. Unlearning False Assumptions
8. The African Builders' Dream Team
9. Learning from Failure

10. How African Builders Define Success beyond Revenue and Conventional Metrics
11. Building for the Future

Building with the APR Framework™

1. Building for Digital Access

> **Question 1.1:** You must launch a critical feature in a market with poor connectivity and high data costs. What is one way you adopted Infrastructure Agnosticism, and what was the quantifiable result (e.g., app size reduction, offline mode, task completion time)?

Anukriti Kedia: A Case Study on Infrastructure Agnosticism and Engineering for Adverse Connectivity

In the markets we serve, most customers use low-end Android phones, juggle expensive data plans, and live in areas with limited network connectivity. This combination creates a fragile user journey—where even the most basic action, like checking an account balance, can feel frustratingly slow and unreliable. Yet for our users on a fintech app, knowing their balance is often a prerequisite to conducting any transaction. If that single action fails, the entire experience collapses.

To address this, we adopted an offline-first design principle and united design, product, and engineering around one shared mandate: to make resilience the default experience. Instead of assuming connectivity, we assumed disconnection. We began caching essential data like balances and user details directly on the device, enabling users to continue transacting even when the network failed. But just providing cached data wasn't enough; it was equally important to accompany the cached piece of information with a timestamp; "Last updated at 3:15 PM", a subtle but powerful gesture that helped us balance transparency with trust and control. They could instantly see how current their data was, transforming uncertainty into confidence.

We revisited our API architecture from the ground up. Redundant calls were eliminated, heavy responses compressed, and background refresh intervals intelligently spaced. What began as 20 API calls for a user flow was reduced to just 1, cutting network and data dependency significantly. From a user experience standpoint, we also tackled the invisible moments, the gaps between intention and confirmation. When data couldn't load, we replaced static spinners with contextual fallback states that clearly communicated what was happening. These small design details transformed waiting from a moment of doubt into a moment of assurance. This cut down our network dependency dramatically, reduced data consumption, and improved perceived loading time on low-bandwidth connections.

This journey taught us that Infrastructure Agnosticism is not simply about building for low bandwidth; it is about respecting unpredictability as a permanent design condition. In environments where connectivity, power, and cost fluctuate daily, resilience becomes the ultimate feature. Designing for these realities doesn't just make products technically efficient; it makes them emotionally dependable. Even when the world goes offline, the experience must remain online, steady, and human.

Question 1.2: Describe a specific challenge where you had to strip back a feature (such as removing a beautiful animation banner or lowering image quality) to prioritize data efficiency. How do you balance the need for a visually appealing, trustworthy product with the user's need to save their precious mobile data?

Adeola Adeyemi: A Case Study on Designing for Low Data Environments

When I joined M-KOPA as PM for the Customer Interactions team, I discovered that about 80% of customer complaints were about *"Phone Unlocking"*—customers struggled to unlock their financed devices after making payments. M-KOPA's core business is device financing, where customers in five African countries pay off phones in installments until full ownership. I found that the customer app had an embedded YouTube video explaining how to unlock a phone, but it wasn't solving the problem.

Data revealed two key issues:

1. Customers rarely played the video due to limited data access
2. Locked devices couldn't even open the app to view it

To prioritize data efficiency and accessibility, we removed the embedded video and replaced it with a lightweight SMS-based solution outlining simple unlocking steps. We also worked with the engineering team to improve the unlocking technology itself.

This change not only reduced data dependency but significantly improved customer success rates in unlocking devices—balancing a simpler user experience with real constraints in our markets.

Adewunmi Aladenusi: A Case Study on Pivoting from Visual Aesthetics to Efficiency

I used to prioritize visually stunning interfaces, believing that a well-crafted design was paramount. However, a significant turning point occurred during a Flutterwave project. We launched a new feature with rich imagery for merchant dashboards that, while elegant on a designer's screen, consumed excessive data on budget Android devices.

Subtle complaints began trickling in from our users, expressing that while our app was "nice," their data ran out too quickly. This feedback wasn't angry, but rather conveyed fatigue, making me realize we were inadvertently excluding the very users we aimed to serve.

In response, we streamlined the experience. We compressed images, replaced real-time animations with lightweight motion cues. The result was less flashy but significantly faster, demonstrating respect for our users' realities.

I've since learned that true beauty in design lies in "effortless efficiency." For an African audience, this means balancing aspirational design with empathy. Users desire engaging products, but not at the expense of performance. A beautiful loading animation that devours 10MB becomes a hindrance, not a benefit.

Now, my design process always considers "tiers of access": how will this perform on a ₦50k Android phone with a 3G network? If it fails there, it's not inclusive. The critical question shifts from "how can this look better?" to "how can this be lighter without sacrificing its essence?" Ultimately, the most trustworthy product is one that respects its users' bandwidth.

Mutugi Kirema: A Case Study on Fancy Animations and High Data Costs

In one of my projects back in 2023 for a top-tier East African bank, we initially designed an animated banner for the home screen. It rotated through credit offers and savings prompts, giving the app a sleek, modern feel. However, during user research across Kenya, Uganda, Rwanda, and Tanzania, we quickly learned that what looked beautiful in the design studio wasn't practical in the real world. Many users were on entry-level smartphones with limited memory and expensive mobile data plans. Some even said, "Every second the app loads, I'm counting my bundles."

We ran field tests to measure data usage and app performance under typical network conditions. The banner alone consumed up to 30% of the total data per session. That insight forced us to strip it back completely. We replaced it with a static image and introduced a "Low Data Mode" option that allowed users to turn off rich media altogether. The result? A 33% drop in data consumption and a significant reduction in app abandonment during loading.

The real balance came from understanding that visual trust isn't built solely through animations or high-res imagery. It's also built through clarity, speed, and respect for the user's resources. We retained strong brand colors, simplified icons, and crisp typography to keep the experience elegant yet lightweight. Ultimately, prioritizing efficiency over flair made the product more inclusive.

2. Building for Digital Literacy

Question 2.1: When designing for users with low digital literacy, what common design elements (e.g., swiping, hamburger menus, complex typography) did you find that did not work for your product and its users so you had to eliminate or completely simplify? What did you replace it with?

Adeola Adeyemi: Case Study on Adapting UX Patterns for First-Time Smartphone Users

I worked on a cash loan product for a mixed group of customers—with about 60% of the customer base being people who hadn't used a smartphone before. So, asking them to log in to their app, and apply for a financial product which they really wanted had to be simple, and efficient.

Some design elements that didn't work and how we worked around them:

1. Swiping without an Arrow Indication—This often left the customers confused on what to do on the page. In fact, we gathered through research (using Maze and our cash loan flow) that some customers would spend up to 20s on the page tapping it profusely, and then leave the page completely frustrated because they couldn't simply figure out what the next step was. Instead of swiping actions, we replaced it with simple arrows that indicated that they had to "move" to a next page
2. Colors—Customers associated the color green to mean action, and the color red to mean danger. So, if we needed them to understand CTA better, most of the colors and simple texts helped them to perform the function faster
3. Search Bars—We found that search bars were more useful to our group of customers if they indicated a sample text of what to search. If not, customers wouldn't utilize that function at all
4. Understanding "Language" with Legal Terms—Loan products typically come with a legal section that requires consent—A simple indicative button that allows customers to click "I accept" or "I deny." A small chunk of our customers feared the "language" as being too formal, and that they weren't fully aware of what they were consenting to even though the details for the T&C's were fully present. Another observation is that customers do not like to read. We improved this by adjusting the language with some consent from the legal team and also indicating the

same variation of the language with translations. We also added some more context into how that section of the loan application was for their protection—these eased customers into proceeding

Paul Otu: A Case Study on Literacy Scalability and User Motivation

We were building a hardware-and-app consumer payment collection solution for lower-income communities. We assumed lower digital literacy and thought we had to build the most basic, barebones interface possible.

During our research, we observed what apps our target users actually had on their phones. We were shocked to see that these "low digital literacy" users were confidently navigating some of the most complex apps in the market, like microloan apps. Apps that even we, as designers, sometimes find confusing. These platforms have objectively complex user experiences—multiple navigation layers, confusing odds displays, dense information architecture, and complex multi-step workflows. By every principle of good design we'd been taught, these platforms should have failed spectacularly with our target demographic. They'd taught themselves, created mental models, found workarounds, and mastered complexity we'd assumed was beyond them.

We realized that the real challenge was motivation. When the incentive is compelling enough, when there's real money on the line, clear benefit waiting, users will climb any learning curve. This changed how we approached design. We stopped asking "How simple can we make this?" and started asking "Is our value proposition strong enough to justify the cognitive effort?" For high-stake features with clear monetary or social rewards? We could afford thoughtful complexity if it delivered meaningfully better outcomes. For low-stakes features? Brutal simplification remained essential. For everything else? Progressive disclosure—start simple, reveal power gradually as users engage.

Respect your users' intelligence. Don't underestimate them based on literacy proxies. The same person you think can't handle a three-step flow is actively managing complex betting calculations in their head. Make the benefit irresistible. They'll figure out the interface themselves. That's the real insight about Literacy Scalability in African markets—it's not only about capability but also motivation and desirability.

3. Building for Socio-Cultural Context

Question 3.1: Beyond translating text, give one clear example of Hyper-Localization that your product achieved. For example, you have to adapt your visual design (color, imagery) or your marketing strategy to honor local social norms? How did this impact user trust?—Building for Socio-Cultural Context.

Ayomide Ekerin: A Case Study on Localization beyond Translations by Adapting Imagery

The thing with running campaigns or launching a product in some parts of Northern Nigeria, for instance, is that it's not just about showing up with your product and a few Hausa translations. Up there, trust and acceptance are built through people, especially religious and traditional authority figures who hold deep influence in their communities. You can have a good platform, the fastest onboarding, the slickest UI, but if religious leaders, community heads, or market elders haven't referenced you directly or through usage, your product may not get the kind of grip you are looking for, even with some level of virality.

So, when we launched our banking app in those areas, we didn't just translate copy; we adapted our entire approach to reflect the community's values and social structure. We worked with local creatives and cultural advisors to ensure our imagery felt familiar—modestly dressed people, everyday Northern scenes, local markets, and subtle use of colors that align with cultural preferences (earth tones, greens, and neutrals rather than loud, flashy hues). Even our photo direction mattered, ensuring men and women were represented in respectful, culturally sensitive ways.

Another game changer was leveraging trusted community voices. We engaged local religious and community leaders who also serve as informal financial guides for their people. This kind of hyper-localization went beyond marketing for us. It was more about embedding trust in the places where trust naturally lives.

Tochy Emereole: A Case Study on Earning Credibility through Cultural Intelligence and Hyper-Localization

Beyond text translation, one strong example of hyper-localization in my work came during my time at IM Mastery Academy. We developed a billboard campaign for Johannesburg's multicultural market, and one of the key headlines included a word that, while harmless in English, carried a negative cultural connotation for certain South African communities.

Given Johannesburg's diverse demographic, we immediately pulled back to redesign and that moment reinforced for me how deeply language, culture, and symbolism are intertwined in African markets. We reworked the copy by collaborating with local linguists and cultural consultants to identify terminology that was universally positive across Zulu, Xhosa, Afrikaans, and English-speaking audiences. We also adjusted imagery to reflect Johannesburg's multicultural identity, ensuring representation that resonated across communities.

This intervention prevented potential backlash and demonstrated cultural respect. Post-campaign surveys showed a 34% higher trust score in the South African market compared to our non-localized campaigns in other regions, with specific feedback citing "felt like it was made for us as against just being translated."

Since that experience in 2021, I've made cultural intelligence a cornerstone of my marketing approach. Whether it's understanding what colors mean across regions, how typography signals tone, or which local expressions evoke pride or discomfort, I always embed these nuances into both strategy and creative execution.

At my current workplace, this approach has evolved into a defined storytelling rule: every piece of content targeting the Nigerian market must reflect the Nigerian lived experience; from the chaos of Lagos traffic and the hawkers weaving through cars, to the comfort of a local buka like Iyaba Sara's. These culturally rooted narratives guide not only our copy but also our visual design, ensuring that imagery, tone, and message feel authentically Nigerian.

The impact has been tangible: users consistently respond with higher engagement and trust because the content feels like it was made for them, not just translated to them. Hyper-localization, for me, isn't a tactic; it's how you earn emotional credibility with your audience.

Question 3.2: Describe the single biggest cultural or social behavior (e.g., cash handling, shared devices, collective decision-making) that forced you to change your product's core design. How did you discover this behavior? How did you handle this?

Joshua Adenuga: A Case Study on Aligning Digital Product Flows with Real User Behavior

While developing CivisHealth, an AI-powered Health Management Information System (HMIS) for hospitals and clinics across Nigeria, the single biggest cultural behavior that prompted us to change our product design was the widespread sharing of computers among healthcare staff, combined with the practice of delaying patient documentation until much later in the day.

During field visits and pilot deployments, we observed that most hospitals had only a few desktop computers for entire wards. Doctors used them collectively for consultations, billing, and reporting. During ward rounds, they left the systems behind, took handwritten notes, and later returned to type summaries into the system. Due to heavy workloads and long outpatient queues, many entries were shortened or skipped entirely.

At first, we thought this was simply an infrastructure limitation. However, after shadowing clinicians and interviewing hospital administrators, we realized that it was a cultural pattern built around shared access and time pressure, and that it had serious implications. Delayed or incomplete entries led to inaccurate patient records, revenue leakages, and frequent audit disputes between hospitals and Health Maintenance Organizations (HMOs). In other systems, staff could even edit consultation data days later, often after claims had already been submitted, creating mismatched records that HMOs refused to honor.

To resolve this, we made a major product decision: we extended CivisHealth to mobile devices so that each healthcare worker could log in securely from their own phone using a personal ID. This allowed doctors and nurses to capture consultation notes, vitals, and prescriptions in real time during out-patient engagements and ward rounds. It eliminated their dependence on shared desktops and made documentation part of their normal workflow.

At the same time, we introduced strict edit controls to preserve trust and accuracy. Consultation records can only be edited within 24 hours and only by the original creator. Lab and radiology requests are locked immediately after submission. These measures ensured audit integrity and ended the long-standing conflicts between hospitals and HMOs over altered data or date mismatches.

This change fundamentally shifted staff behavior: documentation became timelier, audit trails remained intact, and HMO claim rejections dropped sharply. Hospital administrators gained confidence in their digital records, and clinicians reported that the system finally "fit how they actually worked."

By addressing this single cultural behavior—shared device use and delayed entry—we didn't just improve usability; we strengthened accountability and trust across the entire healthcare ecosystem. It taught us that in Africa, good product design must often adapt to human patterns before humans can adapt to technology.

Ope Adeyemi: A Case Study on Redesigning Identity Verification for Shared Device Contexts

One of the most significant cultural behaviors that forced a core design change in our onboarding and KYC products was the use of shared devices and collective account setup in informal markets. In many peri-urban and rural communities across Nigeria, a single smartphone is often shared among multiple users, sometimes an entire household, and in business clusters, by several merchants who rely on one device to access financial services.

We discovered this pattern through field research and escalation trends. When we investigated further through merchant interviews and agent ride-alongs, we learned that many customers depended on their business relationship managers (BRMs) to open or manage their accounts using the BRM's own phone. While this behavior was socially convenient, it severely undermined identity integrity, KYC accuracy, and transaction accountability.

We addressed this by redesigning our onboarding architecture to de-couple identity verification from the device while still respecting local realities. First, we introduced device-binding with biometric consent, requiring each customer to verify ownership of a device before critical actions such as login. Second, we implemented multi-device session management, allowing secure logins from different devices without compromising risk controls. Finally, we rolled out face re-verification on suspicious sessions, using our fraud intelligence layer to detect anomalies in login patterns based on signals such as distance from trusted location and transaction activity.

Princess Akari: A Case Study on Bridging the Digital-To-Physical Gap with a Hybrid Cash Deposit Feature

I worked for a digital bank in Nigeria, and our core service was offering digital financial tools/services for businesses. Everything went well for the first one to two years as we targeted certain demographics of people who were digital-first customers (startup founders, freelancers, tech-savvy business owners, etc.), and these users were quite familiar with digital banking. As time went on, we started attracting other user segments that did some volume in transactions (market traders, supermarkets, etc.). We noticed that there was still a heavy usage of cash among these business owners. They receive cash from their customers for the services they render and sometimes use cash to trade.

While our product was designed as a digital-first, we soon realized that the cash-heavy businesses were growing rapidly in our user base, and they were struggling to fit seamlessly into a digital banking flow that assumed most transactions would start electronically. This insight emerged through feedback loops with our on-ground sales agents. These business owners sometimes receive the bulk of their daily income in physical cash and lack a way to deposit it into their digital accounts with us. Traditional banks have physical branches and smaller kiosks where these business owners can deposit their cash after a full day of sales, whereas our product, being fully digital, didn't have that. Rather than trying to change the user behavior to fully digital, we decided to adapt our product to meet them where they were. We built a cash deposit feature that allowed them to walk into a partner bank, deposit their cash, and have it automatically reflected in their digital account with us. This hybrid approach helped us bridge the gap between physical and digital and respected the users' context and existing cash-heavy habits, while still nudging them toward digital adoption. Over time, it helped the users gain confidence in our digital bank because it supported, rather than disrupted, their already established workflows offline.

Bridget Iyere: A Case Study on Adapting to Shared Device Culture in African Households

When we first designed the eLimu World EdTech platform for children aged four to eight, we envisioned a fun and safe digital space where kids could explore, learn, and grow independently. The platform used game-based learning to make early education engaging and accessible, and our first instinct was to make it "child-first." Children could sign up, create a username, and personalize their learning experience, similar to many global children's learning platforms.

However, during early usability tests and pilot programs in Nigeria and Kenya specifically, we quickly discovered a critical cultural and social behavior that challenged this design assumption—device sharing. In many African households, especially in urban middle and lower-income families, mobile phones or tablets are shared resources. Parents, siblings, and sometimes even extended family members use the same device. A single smartphone could serve as both a parent's work tool and a child's learning device.

This behavior had direct implications for our product. Children frequently logged out or lost progress because another family member used the device. Parents were also uneasy about children accessing the internet independently, and our design unintentionally created a barrier to trust. In some households, parents even deleted the app after noticing their child attempting to "sign up" on their own. Moreover, legal and ethical considerations, particularly around child data protection and parental consent, became increasingly important as we planned for scale.

We discovered this behavior through a mix of user research methods: parent interviews and analysis of usage logs. Our data revealed that over 60% of sessions occurred on shared devices, and the churn rate was highest among households where parents were unaware or uninvolved in setup.

Recognizing this, we pivoted from a "child-first" to a parent-first design model. We restructured onboarding so that parents created and managed the main account, with the option to add multiple child profiles under it. Each child's learning journey, progress, preferences, and rewards were tracked individually, but the parent had full oversight.

This change did more than solve a usability problem. It completely transformed how families engaged with the product. Parents became active participants, setting goals, monitoring progress, and even receiving weekly learning summaries. The shared device constraint turned into an opportunity to build family-centered learning moments, where parents and children interacted through the app together.

The results were remarkable. Parental engagement grew significantly, session completion rates increased, and retention improved from roughly 20% to over 50% within three months of launch. Beyond metrics, the emotional connection families built with the product deepened its value and trust.

In hindsight, what seemed like a usability barrier—shared devices was actually a doorway into deeper cultural insight. It taught us that designing for the African market requires not just translating global design practices but rethinking them around how people truly live, share, and connect. By integrating, rather than resisting, local behaviors, we built a product that felt more natural, trusted, and loved in its environment.

4. Building for Trust

Question 4.1: The APR Framework™ emphasizes Trust-Centric Interaction. How does your brand's messaging, imagery, or advertising directly address the deep-seated lack of trust users may have in digital services? Did you leverage physical agents or community leaders in your campaigns to build confidence in the digital product?

Princess Akari: Insights on Bridging the Digital Trust Gap with Human-Led Interactions

Trust is one of the most critical barriers to digital adoption in African markets. Working in a digital bank, a large percentage of customers will not choose your bank for their main bank because they believe you can "crash" at any time, and there is no physical banking hall for them to go and cry if their funds get missing. Many of our customers had opened their first bank accounts through physical interactions with marketers from commercial banks who visited their shops or stalls and helped them sign up and set up their accounts. Making it easy to set up, and usually these banks are quite popular, so they automatically trust them to keep their money safe.

We discovered that trust couldn't be built through branding or marketing alone; it had to be demonstrated through proximity as well. A lot of users in a specific user segment couldn't relate to the ads we had; it was quite sophisticated and aligned with users who were educated and tech savvy. On the other hand, we had users who didn't fit into that persona, but they were eager to use our product and were businesses. These users needed to see and touch the people behind the product before trusting their money to it, especially because it was not a very popular bank.

In response, we created human-led campaigns showcasing the people who work in the company, improved messaging and visuals, showcased real business owners (not models). In addition to that, we also designed and built a dedicated Sales Agent App to support our efforts to be more trustworthy and human. These sales agents were trained to help new customers with onboarding and KYC verification. They would physically visit markets and stores, help traders and business owners capture their photos and their identification documents, and generally help them set up their digital accounts on-site.

This physical-first onboarding approach helped create a human bridge between the digital brand and the end user, allowing us to build credibility and comfort through familiar social structures. The presence of these agents drove sign-ups and also helped, creating a pathway for better acquisition and retention.

Femi Hanson: Insights on Earning Trust through Human Infrastructure

For most Nigerians, money isn't just a medium of exchange—it's emotional. It represents safety, survival, and dignity. That reality shapes everything about how financial technology is built and adopted across Africa.

As one of the early pioneers in Nigeria's fintech revolution, my experience launching and scaling two of the country's biggest digital finance platforms—OPay and PalmPay—taught me that our biggest challenge wasn't *competition* or *product adoption*. It was skepticism.

Users had been burned by failed transactions, frozen accounts, and fraudulent platforms. So, when we first launched in 2018–2019, the average Nigerian didn't trust "digital money." They wanted to see, touch, and test the system before believing in it.

We realized early on that trust could not be built from behind a screen. We had to meet people physically—in their markets, schools, and motor parks. We had to show up where trust is earned in Nigeria: face-to-face. That's how we began building our nationwide distribution rails—a human infrastructure that connected digital banking to everyday life.

Our agent network became our first trust layer, transforming abstract technology into human relationships. These agents, aggregators, and business developers weren't just transaction points; they were brand ambassadors and customer educators. They explained, demonstrated, and validated that every transaction worked.

We complemented this human strategy with representation. Our campaigns featured real Nigerians—traders, drivers, students—not models. We told authentic stories of *resilience, ambition, and progress*, often in local languages. By reflecting the people we served, we earned cultural permission to be part of their lives.

Within a few years, our brand sentiment improved significantly. That transformation wasn't driven by marketing spend—it was powered by visibility, service, and consistency. We didn't just tell people to trust us; we showed them why they could.

The lesson is simple: In Africa, trust is not assumed—it is earned, person by person, transaction by transaction.

Tunde Mason: Insights on Translating Digital Credibility into Tangible UX

I'd say the end-to-end experience of the Kuda card. Since Kuda is a digital-only bank (we launched in 2019), we needed to have something physical for our customers to hold on to believe we were a real bank and not just a "digital wallet." In our early onboarding flows, we gave priority to card requests, so it was top of mind as soon as the customer completed their KYC. We also made sure, as much as possible, that the customer knew where the card was from the moment of ordering it to the moment it got to their doorstep, with the card tracker on the app. We launched without it, but a surge in support tickets about the status of card requests made it a no-brainer to design a tracker. This tracker also sent updates via email and push notifications as the card would get closer to its delivery date.

The packaging the card was delivered in was something we also paid particular attention to. It wasn't delivered in a basic envelope but in a meticulously designed slider that customers loved to share on social media. That validation definitely contributed to building trust in the Kuda product.

Card deliveries were also not very common in Nigeria at the time (2019/2020), so it was a great way to introduce people to what could be termed a "premium" service offering. Convenience wins you trust.

Paul Otu: Insights on the Psychology of "Where Do I Go Scatter?"

Across several product and solution portfolios I have been privileged to work on—from consumer to enterprise solutions (and everything in between), certain themes reign constant—solutions with human-manned touchpoints consistently outperformed purely digital alternatives. Significantly.

We discovered what users were really asking before trusting any digital financial service: "If something goes wrong, where do I go 'scatter'?"[1] Apps with only an email address feel like scams waiting to happen. People want an address they can show up to because they are rationally cautious. When fraud is common and legal recourse is limited, demanding physical accountability is smart consumer behavior.

Even enterprise clients exhibit this behavior. People prefer picking up the phone to speak to "their guy in the system" (might be an account manager, might be some random person they've built rapport with). The self-service portal exists and it works. But when something important happens, they want a human who they can hold accountable. Digital services operating purely in the cloud "feel" like they can vanish overnight because many have. Physical presence signals: "We're real. We're here. We can't run."

This makes sense when you understand African commerce. The traditional thrift collector knows every trader's face, family, and reputation. When the month-end comes, everyone's comfortable. Why? Because of the trust built through repeated human contact.

We can build digital apps with better features and 24/7 access. But if we eliminate the human element entirely, we've broken the mental model that makes financial services trustworthy.

Simple things work: Display an account officer's name and phone number prominently in your app. Send emails from "Wale from Interswitch" instead of noreply@company.com. Show faces of real team members. Give users a human name they can reference when things go wrong.

5. Conducting User Research in Africa

Question 5.1: Describe the single most difficult obstacle you faced when conducting user research (UXR). What unconventional research method or local tool did you invent or adapt to gather essential user data?

Adewunmi Aladenusi: A Case Study on Overcoming User Skepticism through Immersive, In-Person Engagement

In African markets, user research faces a significant hurdle: a pervasive lack of trust. Users are often skeptical due to past negative experiences with unreliable applications and unfulfilled promises.

Researchers are met with immediate scrutiny, facing questions like, "Who are you, and will this truly help me?" This skepticism stems from a history of products that failed to deliver on their promises, leaving users feeling exploited and wary of engaging in any further data collection efforts.

My work at Flutterwave, specifically on the Agent App and POS products, exposed the limitations of conventional research methodologies. To truly grasp what "seamless" meant for merchants operating with unreliable internet and power, direct engagement was crucial. I moved beyond surveys and Zoom calls, opting instead to visit their physical shops and meet them in their environments. This approach was essential for understanding the nuances of their daily operations and the specific challenges they faced.

This immersive approach involved actively assisting merchants with payments, logging sales, and resolving customer issues. This hands-on method yielded invaluable insights, revealing that the fear of a failed transaction transcended financial loss, extending to emotional embarrassment and a damaged reputation within their community. By stepping into their shoes, I was able to observe firsthand the emotional and practical impact of a system that wasn't working as intended.

Ultimately, my key learning was that in Africa, research is not merely about extracting insights; it's about "earning honesty." Once users perceive genuine listening and a sincere effort to understand their needs, rather than just an attempt to exploit their pain points, they readily share valuable and truthful data. This trust-building process is paramount to conducting effective and meaningful research that genuinely addresses the unique challenges and opportunities within these markets.

Mutugi Kirema: A Case Study on "Research by Proxy" to Gain Insights on Sensitive Topics

One of the biggest challenges I've faced in conducting user research in African markets was getting users to speak honestly about money-related mistakes. In one mobile banking study, participants were visibly uneasy when asked about overdrafts, loan defaults, or accidentally sending money to the wrong number.

Many feared that admitting such experiences would make them seem "careless" or "financially irresponsible." I remember one lady saying, "I once borrowed again just to repay the first loan… I don't want that recorded." That kind of hesitation made it difficult to surface real pain points around financial stress and digital literacy.

To navigate this, my team introduced a method we called "research by proxy." Instead of asking directly, "Have you ever defaulted on a loan?" we framed it as, "Do you know anyone who has struggled to repay through the app?" That small shift unlocked powerful stories, users described "friends" or "relatives" who missed repayment reminders, misunderstood interest charges, or avoided customer care due to shame. In reality, they were describing their own experiences, safely disguised.

Through these adaptations, we captured more genuine insights and understood that in African contexts, authentic data emerges when research feels safe and not evaluative. Designing trust into the research process is often the hardest, but most valuable, method of all.

Bridget Iyere: A Case Study on Navigating User Research Challenges with Young Learners

Conducting user research for eLimu EdTech platform, designed for children aged 4 to 8, presented unique challenges. Unlike adults, young children can't easily articulate their needs or feedback through conventional methods like surveys or interviews. This meant we had to get creative and adapt our approach to truly understand their experiences.

One of the primary obstacles was the shared-device environment. Children typically accessed the app on their parents' phones or tablets, making it hard to observe their natural interactions without adult influence. Additionally, children often felt the need to respond in ways they thought adults wanted, making genuine feedback elusive.

To overcome this, we partnered with a sister company that runs physical learning centers, StemCafe. By creating a community of young users, whom we called "eLimu Geniuses", we established trust and rapport. This approach allowed us to observe children in a relaxed, playful environment, where they interacted naturally with the app. We closely observed their facial expressions, body language, and engagement levels at different points in the app, providing insights that we couldn't capture through direct questioning.

An essential part of this process was building a warm and trusting environment. Children needed to feel comfortable and understood. For example, during testing, one child pointed out that our mascot looked "naked," which led us to adjust its design. Another instance involved a character's facial expression that kids perceived as bullying, prompting us to make adjustments to the visuals. These real-time insights from the kids were invaluable in refining our product.

For parents, gathering feedback was more straightforward, often through surveys or direct conversations. But with young children, direct observation and fostering trust were the keys to genuine insights.

In summary, the challenge of researching young learners pushed us to innovate. By combining community-based testing, careful observation, and building rapport, we were able to create a more effective, child-friendly product.

Susan Onigbinde: A Case Study on Improvising Low-Tech Solutions for Contextual Research in Rural Areas during Lockdowns

The most difficult obstacle we faced was conducting contextual user research during the government-imposed COVID lockdown, which made it impossible to meet rural farmers in Kaduna and Plateau states in person. Our firm, based in Lagos, was working with the product team of an agricultural technology company to optimize their digital app, used by farmers and others in the farming ecosystem to improve their output. The lockdown in Lagos at the time was heavily enforced and at the time, we were right in the middle of the project.

We were faced with a number of challenges; however, our main one was understanding the real-life experiences of the farmers, lead farmers, and their supervisors. We needed to deeply understand how they went about their day, and how they used the app in the context of their daily work. Most of these farmers lived in rural communities and were largely illiterate, which made conventional remote methods difficult.

To overcome this, we created a low-tech, WhatsApp-based diary study. We mapped out a typical day for each farmer, based on the journeys of farmers from home to farm to market, and designed simple MS Word templates with prompts for photos, comments, and their reflections. Each section included simple distinct boxes showing where they were to insert specific images or notes; we then shared these templates with supervisors via WhatsApp. These supervisors helped translate and document responses for farmers who couldn't write in English.

This approach worked very well. When the completed diaries and photos started coming in, we were able to see and understand the beans and groundnut farmers' realities without being there physically. Their work and lives around their homes, farms, market interactions, and how they utilized the app as they went through their day, became very clear to us. The materials they sent gave us clear insights into their contexts, routines, and behaviors.

These findings became the foundation for the usability tests we conducted later. We scheduled remote sessions with each farmer and supervisor using their local office laptops, testing design improvements based on what we had already learned.

This somewhat unconventional method, forced on us by the pandemic, provided detailed contextual data that allowed us to make informed design recommendations. It also highlighted an important lesson for me, that meaningful, human-centered research does not always require highly sophisticated tools; sometimes, creativity and local adaptation are just as effective.

Akinyi Odera: A Case Study on Balancing Research and Health Care Service Delivery

Over a four-year project period, ThinkPlace conducted extensive research and design sprints in health centers in remote regions of Ethiopia. The intention was to understand the social and psychological influences on behavior among frontline health workers who manage patient information. The researchers aimed to go beyond the reported data and gain insights into the subconscious influences of data recording and reporting behaviors. The nature of this research required unstructured conversations, observations, and participatory techniques. Researchers realized that the research activities would often be interrupted by the health workers when they needed to attend to patients. This highlighted the need to establish new research routines that would balance insight generation with health care service provision. ThinkPlace developed multiple approaches in response to this unique research context:

- Multiple, Shorter, and Lighter Research Sprints: Breaking down the research schedule into shorter, rapid sprints that allow us to delve deeper into "hot issues" identified during daily rapid synthesis sessions, enabling researchers and health workers to spend time on areas where they are most needed, until saturation
- Spacing out Design Sprints: This gave health workers time to internalize the early solution blueprints, assess their fit into routine workflows, and share their feedback and proposed iteration opportunities, before researchers conducted follow-up activities to guide them on how to embed the early solutions into the health facility's workflows and structures
- Collaboration with Insiders: Researchers established a group of prototype champions in each health facility. These were people identified as informal leaders in the health facility, proactive, and willing to collaborate in supporting the adoption and iteration of the early solutions. The prototype champions-maintained contact with both the health workers and researchers and led any required re-briefing tasks, flagging any new needs or challenges. This progressively built ownership into the insight generation and design work

Question 5.2: How have you adapted or created methods, tools, and so on, given the constraints of digital literacy, building for trust, sociocultural factors, and so on, for more impactful and useful research at all stages of the research process (planning, recruitment, conducting, etc.)?

Yann Le Beux: Insights on Overcoming Cultural Norms to Elicit Honest Feedback in Qualitative Research

Traditional research methods like focus groups and direct individual interviews often fail to yield honest feedback due to social norms. In cultures that prioritize respect and fear giving offense (e.g., Senegal, Ghana, Benin Republic), participants are highly respectful and reluctant to criticize a service, even if they dislike it. This is compounded by power dynamics in group settings where elders' views dominate and the tendency for "individual" interviews to become de facto group sessions with family present. This contrasts with countries like Nigeria, where users are often more outspoken and critical of products and services. The lack of open criticism severely impacts the quality and contextualization of qualitative and quantitative data (e.g., artificially inflated happiness scores on a scale).

We've found Role-Play and Indirect Elicitation techniques to work tremendously well. By asking participants to speak through a character, a friend, or by acting out a scenario, there is emotional distance for the participants. For example, for sensitive topics like teenager online behaviors or financial decision-making, participants would share issues by saying, "I have a friend who did this…" or fully embodying a character in a role-play.

Role-play successfully encourages participants to "open up" and discuss problems, negative experiences, and even potential harms they would otherwise withhold. In high-respect, community-focused cultures, research design must intentionally build a bridge of emotional distance to access true user insights.

Akinyi Odera: A Case Study on Adapting Methods for the Constraints of Digital Literacy, Building Trust, and Sociocultural Factors

Balancing Online and Offline Approaches

While the effects of the digital era are evident in Africa, ThinkPlace's research in Ethiopian health facilities indicated that digital readiness will vary across facilities. Beyond access to foundational infrastructure such as digital devices and the internet, the health workers' openness to interact with digital data devices was a strong learning. Where the goal was to visualize the health facility's performance for a wider audience, there was significant value in maintaining offline tools, such as wall-poster dashboards, in health facilities that did not yet have desktops, TV screens, or projectors. Where the goal was to enhance access to skill-building resources, there was significant value in including downloadable, offline, and shareable toolkits as attachments to the digital learning tools. Where the goal was to ensure health workers could easily access data via Telegram or WhatsApp, there was significant value in providing smartphones and data bundles, rather than relying on their personal resources for work-related tasks.

User-Driven Adaptation of Digital Solutions

Building frontline health workers' capacity to adapt solutions to their unique socio-cultural contexts has significantly supported the adoption and sustained use of early solutions. Providing tools and structures to guide self-assessment, for example, contextualized system maps and health facility scenarios, followed by enhancing the health workers' capacity and agency to iterate using approaches such as human-centered design and relevant toolkits, is at the core of "designing with" end users.

6. Navigating Regulations in the African Context

Question 6.1: Point to one specific local regulation (e.g., a central bank rule, a data privacy law) that initially blocked your product launch or marketing campaign. How did your team handle this?

Ope Adeyemi: A Case Study on Leveraging Hybrid Verification to Scale KYC Compliance

A major regulatory challenge we faced was the Central Bank of Nigeria's KYC requirement mandating physical address verification before a customer could be upgraded to Tier 3. This is the level that enables higher transaction limits and broader financial access.

The rule, which was designed to enhance traceability and curb fraud, created a growth bottleneck. Purely physical verification was operationally heavy, costly, and slow especially across 36 states where street addressing systems are inconsistent. It initially limited our ability to scale Tier 3 upgrades without breaching compliance.

We responded by building a hybrid address verification system that merged digital geolocation intelligence with field validation. Customers first confirmed their address digitally by dropping a pin on a map or sharing geolocation data from their mobile device. We then integrated a service provider capable of verifying addresses remotely using location and behavioral signals thereby enabling instant verification when confidence thresholds were met.

For edge cases or low-confidence results, we triggered in-person verification through our Address Verification Officers and BRMs. These agents validated the location by capturing live GPS data, timestamps, and images within a 10-meter radius of the declared address.

This approach fully satisfied CBN's physical verification requirement while dramatically improving efficiency. Verification turnaround time dropped from 5.2 days to under 48 hours, verified Tier 3 accounts grew by over 200% in six months, and compliance accuracy remained at 100% during audits.

Tochy Emereole: A Case Study on Redesigning Compliance to Overcome Growth Bottlenecks

Prior to obtaining our SEC (Securities and Exchange Commission) provisional license in Nigeria, we faced significant constraints on our product launch and marketing strategy for our centralized exchange. We were unable to publicly market certain crypto-related products or run large-scale acquisition campaigns, as doing so risked regulatory violations. This created a bottleneck for user growth and limited the scope of our marketing efforts.

To address this, our product, legal, and marketing teams worked hand in hand to redesign compliance flows, from updating KYC and transaction reporting features to ensuring marketing copy and creatives strictly adhered to SEC communication standards. We also shifted our messaging to emphasize financial literacy, transparency, and user protection, aligning with the regulator's expectations while still resonating with our audience.

The result was transformative. Upon receiving the SEC provisional license, we became the first centralized exchange in Nigeria to achieve this milestone. Following the license announcement, customer research showed a clear pattern: trust and perceived security became the top reasons users chose our platform. Consequently, we saw a significant surge in new user sign-ups and engagement, proving that regulatory compliance, when communicated effectively can be a growth lever, not a limitation.

7. Unlearning False Assumptions

Question 7.1: Before launching your product in an African market, what was the biggest false assumption your team held about the user or the infrastructure? How quickly did you learn the truth, and what immediate product change did you have to make?

Ope Adeyemi: A Case Study on Designing Seamless Onboarding in the Face of Downtimes

Before launching our onboarding platform at Moniepoint, the biggest false assumption we held was that government and third-party infrastructure would be consistently reliable especially for critical APIs like the National Identity Management Commission (NIMC) used for real-time NIN verification. We assumed verification services would perform at scale, allowing us to keep all user onboarding and authentication processes fully online.

We quickly learned how flawed that assumption was. Within weeks of rollout, repeated NIMC downtimes and network disruptions brought onboarding to a standstill. Thousands of legitimate customers were unable to open accounts simply because a government service was unreachable.

Our immediate response was to adopt an infrastructure-agnostic design philosophy. We re-architected the onboarding flow to degrade gracefully when external dependencies failed. For example, we introduced asynchronous verification, allowing customers to continue onboarding even when NIN verification was delayed, with background retries once connectivity was restored.

We also expanded our design to support offline login for biometric-enabled devices, enabling merchants in low-connectivity regions to access their accounts securely without an active internet. In parallel, we added dashboard widget data caching, so users could view balances and recent transactions while offline or during network instability.

Anukriti Kedia: A Case Study on Replacing Minimalist Design with Functional Maximalism

Like many design teams influenced by global standards, we began with the assumption that Western ideals of minimalism: streamlined interfaces, deep navigation hierarchies, and tightly grouped information would translate seamlessly to African contexts. We equated simplicity with usability and elegance with effectiveness.

In practice, the opposite was true. Through usability testing and live feedback, we quickly discovered that many users felt lost inside the app, often unable to locate core functions or even realize they existed in the first place. For users new to digital financial tools, abstraction didn't reduce cognitive load, it increased it. They didn't want to explore or guess. They wanted clarity. They wanted to see everything they could do without worrying about what might be hidden.

This realization led us to rethink our entire approach. We flattened the navigation hierarchy, brought key functions to the surface, and made high-frequency actions immediately visible. We paired icons with clear text labels, creating affordances that worked across literacy levels and languages. However, simplifying structure didn't mean showing everything at once. Removing nesting wasn't about putting every feature in plain sight; it was about revealing the right things at the right time, balancing visibility with intention. We also introduced customization, allowing users to reorder or pin the tools most relevant to their daily use.

The change wasn't cosmetic, it was cultural. We stopped designing for the idealized "global" user and started designing for the lived reality of our own. Our goal shifted from aesthetically minimal to cognitively minimal; reducing friction not through fewer elements, but through clearer pathways.

This experience taught us a valuable truth about designing for African markets: usability isn't universal. Patterns that work elsewhere can fail spectacularly here; not because users are behind, but because their digital context is different. Many are multitasking, using shared devices, or operating under pressure and uncertainty. In those conditions, the most respectful interface is one that's explicit, transparent, and easy to trust.

Sometimes, that means replacing minimalist purity with functional maximalism and rediscovering that, in our context, clarity is the new simplicity.

8. The African Builders' Dream Team

Question 8.1: If you could only hire one local expert to ensure your product's success in a new African market, what job title would that person hold (e.g., UX researcher, agent network manager, compliance officer)? What specific knowledge or skill do they possess that your external team could never replicate?

Princess Akari: A Combined Market Research and Ecosystem Advisor Role

If I could only hire one local expert that person would hold a hybrid role between a Market Researcher and a Compliance/Ecosystem Advisor; I don't know what the title of the role might be. Essentially, someone who bridges deep contextual research with regulatory and operational know-how.

I'd be looking for someone who either grew up in that market or has direct lived experience there or has launched a similar product there. Their greatest asset would be an intimate understanding of how the market works on the ground, not just based on "paper."

In African markets, success rarely depends solely on the product's features; it heavily depends on contextual and cultural fit. You would also need to know the informal systems that coexist with formal structures, the social hierarchies that influence trust, and the rules that determine user adoption. A local expert can provide early warnings and direction that no external team could replicate. They would tell you, "This approach won't work here because people prefer this to that," or "You'll need to engage specific local associations before launching in that region." These nuances can save months of trial and error and prevent waste and costly missteps. Africa is a very delicate place, and these cultural contexts matter a lot. Even if you want to use desk research, most of the time the data is not readily available for less popular markets, or they are available but outdated. You almost always need to have a local expert on your expansion team to get a strong understanding of the needs/opportunities, contexts, and barriers.

Also, in many African countries, policy interpretation can differ from one regulator's office to another, and this role would help us navigate that variability strategically.

Ultimately, this local expert would act as a language translator if and where necessary and also help with cultural and operational translations to help connect the external team's product vision to the realities of the local market. They would know how to test ideas with real users, identify trusted local community partners, and flag potential friction points before they become real problems.

Ayomide Ekerin: The Critical Role of a Local Behavioral Insights Manager

If I could only hire one local expert to ensure our product's success in a new African market, it would be for a role I would call "Local Behavioral Insights Manager." Like they say, "there may be a gap in the market, but there may not be a market in the gap," and a lot of times, it's those of us in product marketing who take the fall for a dissonance in market fit.

See, the thing about entering any new market in Africa is that usage behavior doesn't always follow logic, at least not the kind you can observe from a distance. People don't just choose products; they choose what feels familiar, what others around them use, or what fits into the rhythm of their daily life. A behavioral insights manager would live close enough to those realities to understand both the direct and non-direct alternatives our target users already rely on, whether that's a community savings group, an airtime reseller who doubles as a "bank," or a mobile money workaround built on trust, not tech.

This role won't be just about research; it's about decoding human patterns—how trust is built, how convenience is perceived, and what emotional triggers drive adoption. That insight would guide how we tweak an existing product to fit the local gap, maybe it's simplifying KYC, maybe it's redesigning onboarding flows, or maybe it's rethinking our incentive structure entirely. Simply put, this kind of hire would help us figure out the why behind the how: why people behave the way they do, and how we can meet them where they are without forcing them to change too much, too fast. That's the kind of intelligence no external team can replicate.

Tunde Mason: A Local Market UX Researcher Is Essential for Product Success

A UX Researcher with hands-on field experience—ideally someone based in the target country or able to fully integrate into its culture—would be the most critical hire. More specifically, a Local Market UX Researcher who brings deep cultural understanding, language familiarity, and lived insight into the everyday realities of the people we want to serve. It's extremely difficult to truly understand a market from a distance. You need to be present, observing, listening, and absorbing. That immersion is what uncovers real user motivations, fears, habits, and expectations—insights you can only get directly from the source.

African markets are far from uniform. Kenya's mobile money culture is not the same as Nigeria's. South Africans think about onboarding differently than Ugandans. Even within one country, Lagos users behave differently from those in Abuja or Kano. A skilled UX researcher becomes the lens that helps a product team navigate this complexity, rather than assuming one solution fits all.

That said, I don't believe any single role guarantees success. What truly matters is understanding the market you want to enter and confirming that real demand exists for what you're offering. Success also means knowing when to step back—after doing the research and being honest enough to see that your product may not fit, or that you don't yet have what it takes to win in that environment.

Tochy Emereole: The Need for a UX Researcher with Deep Cultural Ethnography Skills

If I could only hire one local expert, it would be a **UX Researcher** but not just any researcher. I need someone who combines rigorous research methodology with deep cultural fluency and on-the-ground access that no external team can replicate.

I refuse to build marketing strategies on social media speculation, competitor assumptions, or surface-level insights. Every campaign I launch must be rooted in empirical, locally sourced data that reveals:

- Actual user behavior, not what we think users want
- Cultural decision-making frameworks that influence product adoption
- Unspoken barriers to trust that wouldn't surface in standard surveys
- Local context and nuance that shapes how people interact with digital financial services

In diverse African markets, assumptions can be costly, consumer behavior often varies not just from country to country, but even from city to city. A local UX researcher brings an irreplaceable depth of on-ground insight: how people think, what drives their choices, what pain points shape their decisions, and what cultural or social factors influence their trust in digital products.

While external teams can analyze trends or social media sentiment, only a local UX researcher can translate lived experiences into actionable data, uncovering truths that aren't visible in analytics dashboards. Their fieldwork, interviews, and usability testing ensure that every product feature, message, and campaign is rooted in how the market actually behaves, not how we assume it should.

In essence, they help me replace speculation with evidence-backed empathy, guiding marketing strategy and product design in ways that truly resonate with the people we want to serve.

I let the market guide my campaigns, not the other way around. A local UX researcher is the bridge between our ambitions and market reality, they ensure we're building trust through relevance, not hoping for it through volume. Without this foundation of real, on-ground data, we're just guessing in an expensive, unfamiliar market.

9. Learning from Failure

Question 9.1: Tell us about a product feature or solution that failed spectacularly in the African market due to a cultural blind spot or an infrastructure assumption you made. What was the single most valuable, non-obvious lesson you learned from that failure that guides your team's design or marketing philosophy today?

Bridget Iyere: A Case Study on Overcoming Cultural and Infrastructural Blindspots

In our endeavor to create the eLimu Tutoring Platform earlier on, we aimed to revolutionize supplementary education for children in Africa. Our vision encompassed parents, children, and tutors as key stakeholders. We assumed that tutors would be at the forefront, using the platform to enhance classroom learning through gamified content in subjects like math, science, and literacy.

However, as we delved deeper, we uncovered a significant cultural and infrastructural blind spot. Many tutors in the region relied on low-end smartphones with limited storage and unstable internet connectivity. This insight led us to design eLimu primarily as a web app, thinking it would be more accessible and less demanding on devices.

Yet, this assumption had unintended consequences. While tutors found the web app convenient, parents and children faced challenges. The open nature of web apps meant that children could easily navigate away to unintended content, raising safety concerns. Parents also struggled with remembering URLs and passwords, which negatively impacted engagement and retention.

The most valuable lesson we learned was the critical importance of understanding the end-user environment. We recognized that, in the African context, parents and children needed a more secure and controlled environment, something a mobile app could better provide. By pivoting to a mobile platform, we offered parents and children a safer, more accessible experience. This shift not only improved user engagement but also boosted retention rates, as parents could now monitor their children's learning more effectively.

Today, this experience guides our design and marketing philosophy. We now prioritize the end-user context, ensuring that we consider device capabilities, security concerns, and the practical realities of all of our users. This approach has enabled us to create more effective, sustainable educational solutions.

10. How African Builders Define Success beyond Revenue and Conventional Metrics

Question 10.1: How do you personally define success for your product in an African market? How applicable are conventional global metrics like "Daily Active Users" for your product? Beyond these standard metrics, how else do you measure success?

Joshua Adenuga: Defining Success beyond Engagement

When I think about success for CivisHealth, I don't look at metrics like Daily Active Users or Average Session Time. Those are useful for consumer apps, but hospitals don't measure success that way and neither should we.

For us, success means that a hospital starts to feel the impact of the system in its daily operations. When a doctor spends less time filling out forms and more time with patients, or when a finance officer tells us their claim reconciliation process now takes minutes instead of days, that's what matters.

We measure the things we can directly influence. Our main goal is Time to Value; how quickly a new hospital starts seeing tangible results after going live. That could be faster registration, fewer billing errors, or real-time visibility into performance. The shorter the time, the stronger our product-market fit.

We also track institutional retention, which shows how many hospitals renew or expand their licenses. It's one thing to get adoption; it's another to become part of their backbone. When a hospital doesn't need to decide to keep using CivisHealth because it's already part of how they operate, that's real retention.

We monitor feature adoption too, not just who logs in but which departments actually use new modules like pharmacy, lab, or the AI assistant. It tells us whether we're solving real problems or just adding features for show.

Beyond the numbers, I pay attention to what I call quiet signals of success:

- When administrators trust our reports enough to make staffing or budgeting decisions
- When HMOs stop rejecting claims because the data always reconciles
- When clinicians start saying "let's check Civis" as naturally as they'd say "let's check the file"

In our market, digital transformation isn't about keeping people online. It's about making their work easier, more accurate, and more trusted.

So while global metrics focus on engagement, ours focus on embeddedness, how deeply CivisHealth fits into hospital life and how reliably it delivers value. For me, the real win is when people stop noticing the software altogether because it has simply become part of how care gets delivered.

Adeola Adeyemi: Redefining Retention for FinTech Loan Products

As one who has built products in FinTech in five African countries, success is first determined in two measures:

1. Success to the Customer—The ability to deliver value by solving an existing problem. Making a process/service/product much simpler, efficient, or better
2. Success to the Business—While delivering value to the customer, what impact does it have on the business/company? Success here could look like retention and engagement metrics

It is also important to contextualize based on the product, and sometimes even the market. Whilst working on the loan product, these were some important metrics to consider:

1. Loan Conversion Rates—of the number of customers who qualify for the cash loan, how many of them actually proceed to complete the process and obtain the loan
2. Repeat Customers—how many customers who finish/repay a first loan go on to take a second (or third) one
3. Customer Lifetime Value—how many months/years a customer remains with the business (i.e., customer lifetime) and qualifies for other financial products (besides the loan)
4. Lifetime Loss Rates—% of customers who exit or default or never return for a subsequent loan

In summary: while global metrics like "Daily Active Users" are useful, for a loan/product retention is more about the duration of the financial relationship and repeat borrowing, rather than simply daily log-ins.

Adewunmi Aladenusi: Measuring Success by Changed Stories

For me, true product success is when it seamlessly integrates into daily life, transforming from "a remittance app" into the simple act of "I just send money."

While global metrics like DAUs or MAUs provide initial insights, they don't fully capture the African reality. Connectivity, income patterns, and trust cycles are unique here. A user might not log in daily, but consistent product reliance, such as every other day, signifies loyalty. Therefore, beyond quantitative metrics, I prioritize return intention—the organic return of users even after encountering issues like bad experiences, network problems, or failed payments.

At Flutterwave, we focused on "emotional retention," measured not through dashboards but by the sentiment in user feedback after problem resolution. Hearing users say, "You guys actually responded fast," was a mark of success. In a market where trust is paramount, reliability becomes the ultimate North Star metric.

Ultimately, I measure success not by the number of users, but by the number of stories that change. Whether it's a POS agent in Kaduna sustaining her business due to faster refunds, or a son abroad sending money back home for family upkeep, these are the true metrics of success.

In Africa, success isn't just about scale; it's about sustainability with soul. It's about how many people a product empowers to remain hopeful, independent, and visible. This is the kind of product success that endures beyond mere version updates.

Femi Hanson: Blending Quantitative Metrics with Human Impact

For me, success in an African fintech context is both quantitative and human. I believe in data discipline—metrics like Daily Active Users (DAU), Monthly Active Users (MAU), retention rates, and transaction frequency are essential to understanding real engagement and business performance. These indicators help us stay accountable, track momentum, and make evidence-based decisions.

However, in a market as nuanced as Africa, numbers alone don't define success. The deeper question is: *are we solving real problems for real people?*

At PalmPay, we pair operational KPIs with impact and perception metrics that reveal the full picture of progress:

- National Awareness and Trust Surveys: We regularly measure how consumers perceive PalmPay—their level of trust, recall, and confidence in our platform as a safe, reliable, and rewarding digital bank
- User Impact Assessments: We capture how our products improve lives—whether by helping a market trader manage her daily turnover, a student save toward tuition, or a family access emergency funds when they need it most
- Product Resonance Analysis: Beyond feature launches, we ask if each solution truly addresses local realities—from data affordability to savings discipline and access to financial literacy

The most powerful KPI is transformation—when we see users move from skepticism to daily dependence, or when PalmPay becomes part of how people live and transact confidently every day.

Success, therefore, is a blend of reach, retention, and relevance. If millions use PalmPay not because they were rewarded to, but because they *trust* it and it makes their financial lives smarter, simpler, and safer—that's the ultimate metric.

Mutugi Kirema: Measuring Value by Advocacy

Back in my Agency days, I led a research project for a digital lending company in Kenya, and we learned that engagement metrics like Daily Active Users were almost meaningless. We noticed that borrowers didn't use the app daily; instead, they used it purposefully. A user might log in just a few times a month, but if they repaid consistently, borrowed again, and even referred others, that was a far stronger signal of trust and satisfaction than any activity count.

Through research, we discovered that financial trust spreads through community influence, not frequency of use. In many towns and villages, people relied on word of mouth, and this meant that one trusted borrower often introduced entire groups to the app. In the rural areas, for instance, a small business owner who had a positive lending experience ended up bringing in nine new users from her local savings group. None of this came from advertising; it came from credibility earned on the ground.

That insight shifted our success framework from "How often are users logging in?" to "Do they believe in this product enough to advocate for it?" We began measuring repeat usage, repayment behavior, and referral-driven adoption as our key indicators of health.

Success in the African Fintech space is about depth of trust, not entirely on volume of clicks. When a user chooses your product again and recommends it to others it signals something no global metric can capture, and that is genuine confidence in your value.

Susan Onigbinde: Tailoring Product Success Metrics to the African Context

Africa's reputation as "virgin territory" has been embraced by the world in two extremes, on one side, the Doubters, and on the other, the Adventurous stakeholders. The latter group has seen vast opportunities and, along with daring African innovators, have found ways to capture real value for African users. They've shown that the uniqueness of the African context, including the industry a product is built for, and the specific people group it targets, demands a tailored set of success metrics for each product.

From our experience building solutions, the key metrics of success often shift depending on the sector, user demographic, and stakeholder goals. This is largely due to contextual factors such as literacy levels, the deeply communal nature of Africa, and the wide spectrum in income levels.

For instance, in finance, when building for High-Net-Worth Individuals (HNIs), metrics such as transactional volume per user, we found, outweighed global metrics like number of active users. We found that digital literacy levels among this group, combined with the cultural deference shown to the affluent, meant that many transactions were performed behind the scenes by bankers and account managers on behalf of their clients. As a result, a metric such as banker activity rate became a more relevant metric of product engagement than traditional DAU or MAU. Further, given the exclusivity of wealth management and private banking features in the product, and the relatively small HNI population in Nigeria, the average transaction size became another closely tracked success metric. Additionally, referral rate emerged as a critical indicator of product trust and satisfaction, reflecting the strong network and relationship-driven behaviors of both this user segment, and the underlying African construct.

In healthcare, however, the dynamics were a bit different. When building for peri-urban and rural, averagely educated African women, success was influenced less by direct usage data and more by social acceptance and community adoption. This group's engagement is shaped by literacy levels and the powerful influence of figures in their lives, in the form of husbands, mothers-in-law, or religious leaders. Gaining that initial trust and approval from these women, that is, level of entry, became a key success metric. Once that was achieved, the communal nature of African women often created a snowball effect, with activities such as word-of-mouth endorsement leading to steady, organic adoption, even at times breaking protocol of seeking the permission of the strong influences in their lives to use the solution.

Ultimately, in African markets, success goes beyond the usual metrics and is defined by how deeply a solution integrates into local realities of the region, divided even further by nuances of gender, societal influence, and social class. Given all these factors, a product thrives by its ability to earn trust, fit into cultural and infrastructural contexts, and create tangible improvements in users' lives.

Question 10.2: Beyond revenue, how does your product address a fundamental societal challenge in your market, such as **financial inclusion**, reducing travel time, or increasing digital literacy?

Joshua Adenuga: Eradicating Geographical Barriers to Continuous Patient Care in African Healthcare

Beyond revenue, the most important problem CivisHealth is solving is the geographical barrier in healthcare access. In most parts of Africa, a patient's medical history lives and dies in the hospital where it was created. When that patient moves to another city or facility, their record doesn't follow them. Doctors are forced to start again from scratch, often with little context or visibility into past diagnoses, allergies, or treatments.

This problem became deeply personal for me when I lost a close friend to cancer. He began his treatment in Akure, but when his condition worsened, we moved him to Lagos for better care. We forgot to bring his test results, and the new hospital had to repeat all the investigations before deciding on the next steps. He didn't survive that delay. That experience made me realize how dangerous disconnected healthcare systems can be. It was one of the moments that inspired my commitment to solving this problem.

With CivisHealth, we are building a system where that kind of loss doesn't have to happen again. The platform gives hospitals secure, real-time access to a patient's complete medical record, vitals, allergies, lab results, and ongoing treatments, no matter where those records were first created. It also supports dependant management, linking family members under one record to reflect how care works in African homes, where health decisions are collective and interconnected.

But CivisHealth isn't just designed to heal patients. It's designed to heal the hospital itself. Our philosophy is simple: "Every healthcare facility deserves to be healthy, not just its patients."

A hospital is a living system. When its internal processes break down—through revenue leakages, duplication, or disjointed workflows—the entire institution suffers. CivisHealth acts as the hospital's digital heartbeat, restoring transparency, harmony, and financial health. It ensures every service is recorded, billed, and reconciled, allowing hospitals to operate efficiently and focus again on what truly matters—patient care.

For me, this is what impact looks like. Not just software adoption or revenue growth, but a future where hospitals across Africa are healthy institutions, connected through data, and patients receive continuous care no matter where they are. When a person walks into any hospital and their medical story follows them instantly, that's not just innovation. That's justice, empathy, and life-saving design in action.

Femi Hanson: Transforming Financial Access into Opportunity and Dignity for Millions in Nigeria

At PalmPay, our mission has always gone beyond transactions—it's about transforming access into opportunity.

A child in Ibadan can now send money instantly to his parents in the village.

A mother in Damboa, Borno State no longer needs to walk miles to a distant bank branch just to withdraw cash.

These may sound like small conveniences today, but just five years ago, they were real barriers that shaped how millions of Nigerians lived and worked.

For us, financial inclusion means dignity—ensuring that geography, education, or income level no longer determine who gets to participate in the digital economy. Today, 1 in 4 PalmPay users confirm that our platform is their first-ever bank account. That single statistic speaks volumes: we're not just digitizing transactions; we're onboarding people into the formal financial system for the first time.

We've also seen a behavioral shift. Through our savings products, people are not just transacting. They're learning to save, budget, and plan. Savings, which once felt abstract or inaccessible, has become a daily habit powered by ease and visibility.

Our impact extends offline, too. We regularly take financial literacy training directly to the people—from the markets of Sabon Gari in Kaduna to Fagge in Kano. There, we train women on budgeting, saving, and business management, while also offering them free health insurance packages through our CSR initiatives.

So, beyond revenue or app downloads, what we're solving are the invisible but powerful challenges of access, trust, and education. We're building a bridge between ambition and opportunity—one transaction, one conversation, one empowered user at a time.

Bridget Iyere: Leveraging Localized and Game-Based Learning to Solve Africa's Foundational Literacy Challenge

At eLimu World, our mission is to provide fun accessible education for curious children across Africa and this goes far beyond generating revenue. We're addressing one of Africa's most pressing challenges: **low foundational literacy and numeracy among young learners**. Across Kenya and Nigeria, millions of children in early grades struggle to read simple texts or understand basic numeracy. This gap affects not only school performance but also long-term social and economic participation.

We began in **Kenya**, building a fun, game-based learning platform for children aged 4–8, aligned with local curricula. By using relatable African stories, familiar environments, and mother-tongue narration, we turned abstract concepts into experiences children understand and enjoy. Learning became play, and play became progress. Parents noticed improved reading confidence, teachers saw better classroom engagement, and children began to associate learning with joy rather than pressure.

When eLimu **expanded into Nigeria**, after being acquired by Co-creation HUB (https://cchub.africa/) we saw the similarities and differences in context. We also localized voices, accents, and storylines to reflect Nigerian culture and classrooms.

Beyond academic outcomes, our work directly advances **digital inclusion**. For many families, eLimu is their first structured digital learning experience. By making the platform mobile-first, low-data, and easy to use, we're helping parents and children gain digital confidence, a skill increasingly vital for economic empowerment in Africa.

Our biggest lesson has been that designing for African learners means **designing for real homes, not idealized ones**. When we build with cultural empathy, understanding shared devices, family involvement, and community learning, education becomes truly inclusive.

Ultimately, our greatest success isn't in revenue figures, but in enabling thousands of children in Kenya, Nigeria, and beyond to **read, reason, and dream confidently**, creating a stronger foundation for Africa's digital future.

Tochy Emereole: Democratizing African Economic Participation with Stablecoin-Based Payment Infrastructure

Our product tackles one of Africa's most persistent barriers to economic participation: the fragmented, expensive, and unreliable nature of cross-border and digital payments that locks millions out of the global economy.

Traditional financial rails in emerging markets are plagued by:

- Transaction Delays: Bank transfers can take three to seven days domestically, and weeks for cross-border payments
- Prohibitive Costs: Remittance fees average 8%–12% in Sub-Saharan Africa, the highest globally
- Chargeback Vulnerability: Merchants lose revenue to fraudulent reversals, making them reluctant to serve certain customer segments
- Limited Interoperability: Payment systems don't communicate across borders, banks, or mobile money providers

These inefficiencies disproportionately impact the unbanked and underbanked; freelancers, small business owners, diaspora families, who are excluded from participating fully in digital commerce.

By building infrastructure around stablecoins and cryptocurrency, we've fundamentally changed the economic equation:

1. Instant Settlement: Transactions complete in seconds. A freelancer in Lagos can receive payment from a client in London and access funds immediately, critical for people living paycheck to paycheck.
2. Dramatically Lower Costs: Transaction fees drop from 8%–12% to under 1%, making micro-transactions economically viable and preserving more value for recipients
3. Irreversible Transactions: Eliminating chargeback risk allows merchants to confidently serve previously "high-risk" customer segments, expanding market access
4. True Interoperability: Stablecoins move seamlessly across borders, platforms, and traditional banking systems, creating a universal payment layer
5. Dollar-Denominated Stability: In markets facing currency volatility, stablecoins provide access to dollar-backed value without requiring foreign bank accounts

Our clients, freelancers, small merchants, and remittance recipients, now participate in the global digital economy on equal footing. A trader in Kano can receive payment from a supplier in Guangzhou as easily as someone in New York can. We're building the financial infrastructure that should have existed all along, democratizing access to fast, affordable, and reliable money movement.

11. Building for the Future

Question 11.1: Imagine the African tech ecosystem in ten years. What is the single biggest infrastructural or social problem your specific field (design, engineering, product) must solve now to fully enable that future vision?

Paul Otu: Building for the Future by Solving Cross-Border Infrastructure

Right now, it's easier for me to do business with someone in London than in Accra. I'm in Lagos. Accra is a few hours away. London is across an ocean. Yet the London transaction is faster, cheaper, more reliable. Solving cross-border infrastructure unlocks everything else Africa needs to build for the next decade.

I've led design for cross-border initiatives at Interswitch—P2P payments, remittance platforms, agency banking across multiple countries. Every single time, we hit the same wall: regulatory frameworks and infrastructure change every "2 kilometers". Different central banks. Different KYC requirements. Different settlement windows. Different everything.

From a design perspective, it is near impossible to create unified experiences when the underlying reality is fragmented chaos.

But imagine a business or product team that has learned to build seamlessly across Lagos, Nairobi, Accra, and Kinshasa? They'll have capabilities few others can match. Organizations that navigate this complexity will build the capacity to serve any complex, heterogeneous market globally.

We just need to stop using Silicon Valley's playbook. We need something new—what I'd like to call the Tiered Localization Model.

- Tier 1: Universal foundations. Identify the core problems that transcend borders. For example, people need to send money. Verify identity. Trust transactions
- Tier 2: Continental backbone. Build vanilla infrastructure that works everywhere, and that localized solutions can run on top of—federated identity standards, interoperable payment rails, common APIs
- Tier 3: Hyper-localization. Same core, contextual/localized delivery. Adapt to frameworks, trust mechanisms, payment behaviors, infrastructure realities, regulations

Solving cross-border trade goes beyond enabling commerce. It also opens the door for innovations in Healthcare (medical records following patients across borders), Agriculture (supply chains that actually work across the continent), Energy (through continental grid optimization), and Talent (seamless hiring and payment across countries)

Every innovation for the next decade will be enabled or constrained by the infrastructure we build today. Luckily, we're not starting from zero. Interswitch processes billions of transactions and has built infrastructure that millions depend on. We just need to think continentally while building locally.

Imagine Africa ten years from now—not 54 fragmented markets, but a unified technological powerhouse with over 1.4 billion people, extraordinary cultural richness, and the fastest-growing economies on Earth.

The diversity that makes this hard is the same diversity that makes us valuable. Markets that master complexity develop capabilities others need. Our challenge is our competitive advantage—if we architect for it properly. Yes, it will take work. Patience. Investment. Policy advocacy. But the opportunity? We're not just building products. We're building the foundation for a continent's digital integration, and I believe that's worth the effort.

Yann Le Beux: Empowering Non-Tech Founders with Local Language AI and Service Design

Broadening the focus from just digital product design to service design could be more helpful in bridging the digital gap. This includes digitizing the people in the value chain of people who can handle technology. We also need to invest in operations, ensuring you have people on the ground to build trust.

The most specific developments expected from AI are speech-to-text and text-to-speech features, which, if perfected, would be a gamechanger by allowing people to communicate through voice with an interface in their own local languages. This requires improved data collection to build models that work with marginalized and low-resource languages. Additionally, I believe that AI will allow more non-tech founders (like traders) with local knowledge to build companies, breaking down the supremacy of tech founders.

Note

1. (In Nigerian parlance, "scatter" in this context means to cause a scene, demand accountability, make noise until your problem is resolved. Works like magic.)

Glossary

Accessibility: The principle and practice of ensuring that digital products and services can be used by everyone, regardless of disability, environment, or personal circumstances. This encompasses physical disabilities (visual, auditory, motor), cognitive impairments, and situational factors such as using a small device outdoors in bright sunlight. In the African context, it often overlaps with usability by focusing on text-only modes like USSD and simplified interactions to serve users on older or feature-limited phones.

African Product Resilience (APR) Framework™: A structured methodology for designing, testing, and deploying digital products in the African context, prioritizing robustness and adaptability to extreme environmental constraints. These constraints include high data costs, intermittent power supply, low digital literacy, and reliance on shared/low-specification devices. The framework emphasizes building products that "fail gracefully" and deliver core value despite underlying infrastructure instability.

Coverage Gap: The percentage of the population without mobile broadband coverage. It impacts digital access: people cannot access the internet if they have no coverage. This has been reducing year on year in Africa.

Digital Access: The ability of individuals and communities to fully participate in the digital economy and society. This includes coverage, device types, affordability, relevance, and knowledge (digital literacy) required to effectively use digital tools. In the context of the APR Framework™, it includes ensuring that solutions are built not only with a mobile-first approach considering that many African users are "mobile-only," but also prioritizing infrastructure agnosticism.

Digital Literacy: The knowledge, skills, and understanding required to use digital technology safely, effectively, and critically. This often includes basic device operation, managing mobile data consumption, identifying scams, and navigating user interfaces based on prior experience with computers and the internet.

Digital Native: A person who has grown up in the digital age and is comfortable with technology, often having spent their entire lives surrounded by the internet and digital devices. While often associated with younger generations globally, the definition in the African context may be narrower, referring to those who had continuous exposure to digital tools throughout their primary education or early career.

Hyper-Local Customization: An advanced form of localization that goes beyond simple translation and currency conversion. It involves deeply tailoring a product's user experience, features, imagery, language dialects, and business logic to reflect the specific, nuanced cultural practices, social norms, and market dynamics of a particular city, region, or demographic group within a country.

Infrastructure Agnosticism: A design philosophy where a digital product is built to function reliably and deliver core utility across a wide range of unpredictable and often severely limited network, power, and device conditions. This means designing the application to minimize data usage, allow for offline functionality, handle dropped connections without losing user data, and function smoothly on basic, low-RAM smartphones.

Interactive Voice Response: A technology that lets a computer talk to a person over the phone. It uses recorded voice prompts and the sounds made when you press keys on your phone's keypad (called DTMF tones) to help you navigate a menu or complete a task without speaking to another person.

Next Billion Users: A collective term, introduced by Google and used by global tech and venture capital firms, referring to the vast population of people who are now coming online for the first time, primarily in economies like Africa, Asia, and Latin America. These users typically access the internet primarily via mobile devices and possess unique digital behaviors and expectations due to low digital literacy and infrastructure constraints.

Privacy by Design: An approach to engineering and product development that integrates user privacy considerations into the system from the initial conception phase, rather than treating privacy as an optional add-on feature or post-launch fix. In the African market, this is crucial due to low digital trust and varying data protection regulations across different nations.

Service Design: The activity of planning and organizing people, infrastructure, communication, and material components of a service in order to improve its quality, the interaction between the service provider and customers, and the customer's overall experience.

Service Thinking: A framework that focuses on the end-to-end experience of a user, viewing the product as just one part of a larger, integrated system (or "service") that delivers value. It emphasizes understanding all touchpoints, both digital and physical, that a user encounters.

Usability: A quality attribute of a system that describes how easy it is to use. The internationally recognized definition focuses on the extent to which a product can be used by specified users to achieve specified goals with **effectiveness**, **efficiency**, and **satisfaction** in a specified context of use. In Africa, the definition is heavily weighted toward efficiency (low data usage, few steps) and effectiveness (successfully completing a transaction despite network hurdles).

Usage Gap: This refers to the percentage of people who use the available coverage. Usage is impacted by digital access based on connectivity, affordability, trust, and literacy.

Index

For Product Safety Concerns and Information please contact our EU representative GPSR@taylorandfrancis.com Taylor & Francis Verlag GmbH, Kaufingerstraße 24, 80331 München, Germany

Batch number: 10782559

Printed by Printforce, the Netherlands